U0195017

献给为广东省建筑设计研究院 ADG · 机场设计研究院十年发展历程作出贡献的人们！

To those who contributed to GDADRI · ADG from 2004 to 2014!

十年之外 十年之间

Within and Beyond a Decade

广东省建筑设计研究院 ADG · 机场设计研究院（2004–2014）作品集

Collection of GDADRI ADG - Airport Design Group (2004-2014)

主编：陈雄　　副主编：郭胜　潘勇　周昶　区彤

Editor-in-Chief: Chen Xiong

Associate Editor-in-Chief: Guo Sheng　Pan Yong　Zhou Chang　Ou Tong

编著单位：广东省建筑设计研究院 机场设计研究院 / ADG建筑创作工作室

Presented by The Architectural Design and Research Institute of Guangdong Province(GDADRI) Airport Design Group (ADG) / ADG Architectural Studio

中国建筑工业出版社

Contents 目录

持守技术本源，致力设计创新

—— 寄语ADG·机场院

容柏生　中国工程院院士
中国工程设计大师
广东省建筑设计研究院终身荣誉总工程师

在广东省建筑设计研究院60多年的历史中，广州新白云国际机场一期航站楼项目无疑是有着广泛社会影响力的作品，它是中国民航史上一次性投资最大的项目之一，是广东省当年规模最大、难度最高、功能复杂、设计新颖、技术先进的公共建筑工程项目，它代表着广东省当年最为卓越的建筑业水平，也是中国民航史上一次转场成功的特大型航站楼的范例。作为广州市的重要门户之一，广州新白云国际机场为南来北往的旅客提供了优质的服务。

1998年省建院为了争取参与这个项目的设计，抽调部分技术骨干专门成立了机场设计专组（当时简称机场组）。从1998年参加国际竞赛并最终获得与美国知名机场设计公司合作的设计机会。由此，这个设计团队迈开了探索和创新的脚步，通过这次合作设计，积累了大型公共建筑设计经验。

2004年新白云国际机场建成投入使用后，省建院以机场组的骨干为班底成立了机场设计研究所（后发展为机场院），近年又在机场院成立了省建院的ADG建筑创作工作室。在此后的时间里，继续在大跨度建筑方面探索创新，并从机场航站楼延伸到体育馆、火车站和交通枢纽等，陆续完成了一批有影响力的作品。其中比较突出的有新白云机场一期航站楼扩建、潮汕机场航站楼、广州亚运馆、花都亚运馆、惠州游泳跳水馆、武汉火车站（武广高铁）、深圳机场航站楼交通中心等。同时，ADG·机场院在超高层建筑方面也进行了不少探索，其中重要的代表作品是东莞海德广场，为满足"东莞之门"的城市设计理念，设计方案以酒店和写字楼超高层双塔在高位大跨度连体，为实现独特的建筑造型做出了具有较大难度的创新设计。

值得指出的是，广州亚运馆是2010年广州亚运会唯一新建的主场馆，是该届亚运会的标志性建筑。它是凭借创新的设计理念、独特的建筑体验、独树一帜的标志性与可实施性的最佳平衡等的综合体现，在2007年的国际设计竞赛中以中国原创获胜中标并且成为实施方案。在短短两年中，这个项目完成了从设计到施工的实施，成为广东省院近年最重要的作品之一。ADG·机场院也成为广东省院一支出色的设计团队。

十年之间，ADG·机场院的作品获得了国内外一系列重要奖项，如新白云机场一期工程荣获国家优秀工程设计金奖、首届全国绿色建筑创新奖；广州亚运馆荣获AAA 2014亚洲建筑师协会奖、两岸四地建筑设计大奖、全国建筑设计行业一等奖；新白云国际机场、广州亚运馆和武汉火车站荣获中国"百年百项杰出土木工程"称号。荣誉的背后凝聚了设计团队无数的努力与付出！

在这册ADG·机场院十年作品集里，我们可以看到众多不单设计新颖，而且实施完成度也很高的优质作品，这就体现了全专业设计团队不懈的职业追求，体现了建筑师杰出的设计统筹能力。此外，团队在学术研究方面也取得了不少成果，包括出版专著、发表论文、学术演讲和获得技术专利等。

可以说，在我国快速建设发展的过程中，ADG·机场院一直持守技术本源，致力设计创新。凭借雄厚的技术实力、出色的设计作品、优质的专业服务等，在业界中已享有较高的声誉。

衷心祝愿广东省建院ADG·机场院能百尺竿头更进一步，注重设计原创，注重技术创新，致力绿色环保，以精心设计、至诚服务的团队精神，为业界提供规划、建筑、室内、景观全方位的优质设计服务，为社会为大众贡献更多更好的作品！

1

Commitment to Technical Fundamentals & Design Innovation

- A Few Words to ADG

Rong Bosheng Member of China Engineering Academy
China Engineering Design Master
Lifetime honorary chief engineer of the Architectural Design & Research Institute of Guangdong Province

Terminal 1 of Guangzhou New Baiyun International Airport definitely stands out in GDADRI's 60+ years' history as a highly influential project of profound significance. Involving one of the largest one-time investments ever in China's civil aviation history, it represents the latest and highest level in the building industry at that time in Guangdong due to its unprecedented scale, complexity, design innovation and application of cutting-edge technologies. It sets up an example of one-time operation transfer of mega airport terminal in China, and, as an important gateway to Guangzhou, offers quality services to passengers from China and beyond.

The airport design team was firstly established in 1998 within GDADRI to participate in the international design competition for the aforesaid project. The team, as the final winner of the competition, collaborated with a renowned American airport design firm, and based on the experiences gained thereof, began its own design practices and innovation.

In 2004 when the New Baiyun International Airport was completed and put into use, GDADRI established the Airport Design Research Institute (later developed to Airport Design Group) based on the Airport Design Team and then ADG Architecture Studio in recently years. Since then, ADG continued the design practices and innovation on large-span buildings and has completed a great number of influential projects ranging from airport terminal to gymnasium, railway station and transportation hub. The representative projects include the Expansion of Terminal 1 of Guangzhou New Baiyun International Airport, Terminal of Chaoshan Airport, Guangzhou Asian Games Gymnasium, Dongfeng Gymnasium for Guangzhou Asian Games, Huizhou Jinshan Lake Swimming and Diving Complex, Wuhan Railway Station (Wuhan-Guangzhou High-Speed Railway) and Ground Transportation Center (GTC) of New Terminal Area, Shenzhen Airport. Meanwhile, ADG explored the design for super high-rise with Dongguan Hyde Plaza as the representative project. To present the design concept of "Gateway to Dongguan", the design connects the hotel and office super high-rises via a large span at higher floor and completes a unique building form through highly challenging design innovation.

As another important project for GDADRI in recently years, Guangzhou Asian Games Gymnasium is the only newly built main venue and the representative building for 2010 Guangzhou Asian Games. With the innovative design concept, unique spatial experience and the optimal equilibrium between identity and feasibility, the design proposal submitted by GDADRI won the international design competition in 2007 and was subsequently realized within merely 2 years. This made ADG – Airport Design Group a competent design team under GDADR.

Over the past decade, ADG has won a series of important international and domestic awards, including Gold Medal of National Excellent Engineering Design and The first "National Green Building Innovation Award" for Terminal 1 of New Baiyun International Airport; AAA 2014 ARCASIA Award, Cross-Strait Architectural Design Award and 1st Prize of National Architectural Design Industry for Guangzhou Asian Games Gymnasium; "100 Outstanding Civil Engineering Projects from 1900 to 2010" for New Baiyun International Airport, Guangzhou Asian Games Gymnasium and Wuhan Railway Station. Behind those awards are the numerous team efforts and strong team spirit.

The Collection of ADG - Airport Design Group (2004-2014) presents many design works of high innovation and implementation level, showcasing the professionalism and expertise of an all-specialty design team and the remarkable design coordination competence of the architects. In addition, the team is also fruitful in academic researches, including the publications, academic papers, academic lectures and technical patents.

By now ADG – Airport Design Group has earned a reputation in the industry with its commitment to the technical fundamentals and design innovation, as well as a competent design team, impressive portfolio and quality professional services etc.

I sincerely wish that ADG – Airport Design Group under GDADRI will make greater progress with their commitment to design originality, technological innovation and environmental protection and, through their full-range quality design service covering planning, architecture, interior and landscaping, present more and better design works in the future.

2

1 广州新白云国际机场一号航站楼
Terminal 1, Guangzhou New Baiyun International Airport

2 广州亚运馆
Guangzhou Asian Games Gymnasium

十年之外·十年之间
——ADG·机场院的回顾与展望

陈雄 广东省建筑设计研究院副总建筑师
机场设计研究院院长
ADG建筑创作工作室主任

在2004年8月5日新白云国际机场投入使用之际，广东省院即决定在机场设计专组（简称机场组）的基础上组建机场设计研究所，目标是在机场航站楼等大跨度大空间建筑的设计领域开拓业务。随着团队发展扩大，2009年对外称为广东省院机场设计研究院。2013年在机场院方案组的基础上又成立了广东省院ADG建筑创作工作室。其实ADG即是Airport Design Group的简称，源于"机场组"。当然，也可以是另一层意思，即Architectural Design Group。

今年是新白云机场启用十年，也是机场院成立十年。回顾已经走过的路，这是非常不平凡的十年，值得欣慰的是我们建立了一支出色的设计团队，为社会贡献了一批富有影响力的建筑作品，在业界已经具有一定的知名度。所有的这些成果都来之不易，包含了团队全体员工超过十年的辛勤劳动、艰辛付出和努力拼搏。作为团队的带头人，我与大家一样深深地为之感到自豪！展望未来，我们要以中国一流设计企业为发展目标，正如容柏生院士所殷切寄语我们的，要继续持守技术本源，致力设计创新，坚定地走好自己的路。

1. 回顾思源：前辈榜样与历史机遇

广东省院是一块事业沃土，一代一代传承着前辈们的优良传统。80年代中期我们来到省设计院时，正值改革开放之初，有幸跟随郭怡昌总建筑师等老一辈专家一起工作。印象最深刻的是郭总多年来一直致力于建筑创作，醉心于设计创新，集百家所长，创自己的路。无论设计项目大小，都是呕心沥血地进行创作，精心构思，反复推敲。每有新作，总是以百倍的热情、拼搏的精神和不懈的努力，施展聪明才智，攀登新的高峰。从郭总身上，我看到了一代建筑大师的风采：勤奋、严谨、睿智、谦和，他为我们树立了崇高的榜样。再后来有机会与胡镇中总建筑师和刘荫培副总建筑师等共事，都可以感受到那一辈老建筑师老工程师的敬业精神，以及对年轻人的关心和培养，指导和提携。

时光转到1998年，白云机场业主为新白云机场举行国际竞赛，包括英国福斯特事务所与荷兰机场公司（NACO）联合体、法国巴黎机场公司（ADP）、美国PARSONS公司与美国URS Greiner公司联合体、加拿大B+H建筑事务所等全球知名机场设计公司参加。广东省院的领导班子毅然决定组织技术骨干（即机场组）第一次参加重大国际竞赛，认为这是广东省院一个千载难逢的好机会。通过参加竞赛拓展视野，向国际公司学习，更为重要的是展现我们的综合实力，争取做后续的合作设计。在各专业院老总的指导下，机场组经过三个月的努力拼搏，终于交出了一份令评委和业主都留下深刻印象的竞赛方案。在1998年的金秋十月，广东省院被白云机场业主确定为新机场的中方设计单位，与美国公司合作设计。从此，一个光荣的使命等待广东省建筑设计研究院去完成！

2. 原始积累：新白云机场一期工程

新白云机场一期工程来之不易，全院从多个部门抽调技术骨干70多人组成新的机场设计组，包括建筑、结构、给排水、电气、空调、市政、交通和概预算全部专业，成立了以容柏生院士为首的省院各专业总工程师组成的技术领导小组。2000年全部设计人集中在一起办公，开始了项目设计的攻坚克难，其中2002年8月机场组的总负责及各专业负责在完成设计工作后全部驻新机场现场开展设计服务，直到2004年机场一次转场成功。

1999年开始，我们对美国几乎所有大型机场进行了考察，对亚洲新机场香港机场、新加坡机场、吉隆坡机场和仁川机场进行了重点考察，对国内北京首都机场和浦东机场也进行上门调研，就大跨度大空间建筑的一系列关键技术做了填补空白式的研究。先后掌握了航站楼工艺（航站楼构型、旅客流程、行李流程、联检流程和信息流程）、大跨度建筑的外围护结构技术（大型金属屋面系统、大面积点式玻璃幕墙系统、张拉膜系统）、大跨度结构技术（桁架技术和无檩式箱形压型钢板、三管梭形钢格构人字柱）、异常复杂的石灰岩岩溶发育地区的基础技术，大型公共建筑的能源管理系统（EMS）、高大空间空调技术、消防设计，还有与航站楼相关的专项技术，包括行李处理、标识系统、大空间照明、各种柜台及公共座椅等。

经过这次的中外合作设计，积累了丰富的大型公共建筑设计经验,尤其在大跨度大空间建筑设计的各个相关专业，广东省院完成原始积累而进入了全国先进行列。而且，我们不单在技术上有所突破，在大项目管理及现场服

1

1-2 广州新白云国际机场一号航站楼
Terminal 1, Guangzhou New Baiyun International Airport

2

1

2

3

务方面也积累了丰富的经验，为承接大型复杂公共建筑设计奠定了技术、管理和服务的坚实基础。

3. 实现原创：广州亚运馆

2007年的10月，2010年广州亚运会唯一的主场馆——广州亚运馆（原称：广州亚运城综合体育馆）项目举行国际设计竞赛，广东省院以独立身份报名通过资格预审。就在报名之前我们分别与两家国外著名的大型事务所商谈过合作。寻求中外合作是我们的初衷，由于商务条件无法达成一致而放弃。毕竟在新白云机场一期工程之后，我们在大型复杂公共建筑方面积累了相当的经验，也建立了相当的技术自信，我们不可能满足于以低比例收费去中外合作，更不可能满足于仅仅做施工图设计。结果竞赛委员会选择了8家中外设计机构参赛，其中有大牌国际公司包括英国扎哈·哈迪德、英国奥雅纳、德国GMP，加上法国公司等共6家外国公司与国内设计院联合体，还有一家国内设计院也独立获入选资格。

由我们ADG·机场院代表广东省院参赛，这是一个非常光荣而艰巨的任务。在很长一段时间以来，中国大型复杂公共建筑的国际竞赛绝大部分中标者都是境外公司，我们只能靠拼搏靠创新才可能打开胜利之门。2007年冬天和2008年的初春特别寒冷，那次的南方雪灾依然留在记忆中。大年初三我们就聚集在办公室，继续进行竞赛第二轮的深化方案设计。设计团队士气旺盛，最终交出了完美的答卷。我们以第一轮排名第三而入围优选方案，可是在第二轮交标之后就没有任何后续工作。2008年5月，突然收到消息可能选用我们的方案，正是山重水复，柳暗花明！我们的方案凭借创新设计理念、独特的建筑体验、标志性和可实施性的绝佳平衡，在2007年到2008年的国际设计竞赛中最终以中国原创获胜并且成为实施方案。在接下来仅仅两年的超短工期里，从设计到制作到施工全部都高完成度实施，成为广州亚运会最为夺目的标志性体育场馆。这是中国建筑师少有的在重大项目设计国际竞赛原创中标实施的案例。

4. 合作共赢：广州科学城公寓

2006年广州科学城科技人员公寓项目举行国际竞赛，我们与日本株式会社佐藤综合计画联合参加，竞赛组织方要求报名单位提供概念方案竞争入围，日方提出了围绕用地以超长体量板楼与纺锤形塔楼相组合的布局概念，并做了工作模型表达在文本之中，这种独特的概念及其表达方式显然打动了评委，我们联合体顺利拿到竞赛仅有的三个名额中的一个。

在接下来的正式投标中，中外双方开展了全过程的深入合作。日方继续深化布局和造型，我们重点负责平面及户型、各专业系统及说明，设计表达及排版结合双方意见，模型公司由双方共同选定，具体工作由我们跟进。整个竞赛过程双方密切合作，彼此尊重，最终的成果融汇了双方的智慧，以绝对优势顺利中标。

项目的实施过程中由于业主更换用地而重新调整方案，但建筑的基本元素及特质仍然维持不变，还是板楼和纺锤形塔楼的组合，并以此新方案做技术设计实施。在立面材料的选择上日方一直给予意见，双方在细部上都精益求精，作品完成度也很高，与当初效果图非常接近。

2010年项目落成，马上成为广州萝岗区的地标建筑。值得指出的是，项目既不是特殊功能的建筑，也没有采用复杂的技术和昂贵的材料，外墙用了涂料和玻璃，阳台用竖条铝合金型材，屋面局部用了钢结构造型，建筑简洁大方却富有韵味，受到业主和社会各方面的充分肯定。

这个作品登上了日本《新建筑》杂志，还作为封面作品登上了中国《建筑学报》，受到两个国家主要建筑专业媒体的认可。由于中外双方在设计全过程合作非常愉快，几年来继续一起参加了多个国际竞赛，并在2011年珠海横琴发展大厦及其周边城市设计的国际竞赛中再次中标。

5. 多元发展：公建与地产并行

机场所成立时主要的目标是重点发展大跨度大空间建筑，走专业化的道路。事实上，我们确实花了很多资源在机场、火车站、体育馆和会展建筑的竞赛和投标。先后参加了昆明机场、长沙机场、南宁机场、潮汕机场、重庆机场、深圳机场、武汉机场、厦门机场、烟台机场、海口机场、浦东机场等十多个机场的投标，参加了亚运会的几个场馆竞赛。继续在大跨度建筑方面探索创新，并从机场航站楼延伸到体育馆、火车站和交通枢纽等，陆续完成了一批有影响力的作品。其中比较突出的有新白云机场一期航站楼扩建、潮汕机场航站楼、广州亚运馆、花都亚运馆、惠州游泳跳水馆、武汉火车站（武广高铁）、深圳机场航站楼交通中心等。

我们在超高层建筑方面也进行了不少探索。2005年中标实施的东莞海德广场是第一个民营发展商的项目，酒店和写字楼超高层双塔高位连体在技术上是非常富有挑战性的。2009年开始我们明显增加了对地产的关注，希望寻找更多项目、更多市场、更多可持续的业务，自然想到了发展商。设计的原理总是相通的，过往在公共建筑积累的技术经验以及服务意识可以借鉴到地产项目，尽管两种类型各有特点。终于，在香港新福港地产公司的商务投标中赢得佛山魁奇路商业综合体项目，并在新福港地产另一个项目广州萝岗区线坑村的竞赛中原创中标。我们先后和东莞康帝酒店集团、东莞新世纪地产、香港新福港地产、大横琴公司、保利地产、恒大地产、利海地产、悦榕庄酒店集团进行了合作。

可以说，公建与地产、原创与合作、大跨度和高层构成了我们多元的业务结构。

6. 注重科研：面向实施的学术研究

我们在完成一系列重要作品的同时，结合业务特点和利用项目平台，策划和进行了面向实施的多种形式的学术研究，包括论文、专著、专利、论坛、评审、授课、出访，在理论上总结提高，开展对外重点宣传，寻求更高的附加值，也对员工在实践与理论结合全面发展上提供机会。

2003年，在新白云机场一期航站楼接近完成时就着手撰写论文，2004年8月通航，《建筑学报》9月作为封面项目刊登了《广州新白云国际机场航站楼》的专题论文，并策划编辑出版航站楼专著，经过两年的不懈努力，终于在2006年由中国建筑工业出版社出版了《广州新白云国际机场一期航站楼》的专著。在2006年《建筑学报》刊登了论文《新白云机场的规划与发展》，在2008年《建筑学报》刊登了论文《机场航站楼发展趋势及设计研究》。机场方面的主要研究成果还包括在2014年《建筑学报》刊登了《干

1-3 广州亚运馆
Guangzhou Asian Games Gymnasium

4 广州科学城科技人员公寓
Scientists' Apartment, Guangzhou Science City

5 揭阳潮汕机场航站楼及配套工程
Jieyang Chaoshan Airport Terminal and Supporting Works

4

5

1

2

线机场航站楼创新实践——潮汕机场航站楼设计》。围绕机场航站楼专题研究成果还刊登在《AT建筑技艺》、《a+a建筑知识》等有影响力的专业媒体。2011年，《建筑学报》2月也是作为封面项目刊登了广州科学城科技人员公寓的相关论文，同年8月日本《新建筑》也刊登了这个项目。

2010年，广州亚运馆建成并在亚运会大放异彩，受到社会各界及主要媒体的重点关注，在10月举办广州亚运会的同时，就在《建筑学报》作为封面项目刊登了《广州亚运馆设计》的专题论文。目前正在编辑出版该项目的专著。

结构专业以广州亚运馆项目为平台开展了一系列卓有成效的科研活动。在《建筑结构学报》发表了代表论文《广州亚运城体操馆结构设计》、《广州亚运城历史展览馆结构设计》、《广州亚运城历史展览馆结构设计》，在《建筑结构》发表了论文《钢结构节点研究》、《拉索幕墙在建筑工程中的应用》等。组织研究并取得了三项发明专利：《蒙皮局部应用于桁架组合结构之安全性的分析测算方法》、《一种有利于消减钢构件节点应力的型钢构件》、《一种可提高抗震能力的钢筋混凝土基础》。

在十年间，我们共发表在各专业核心杂志的重要论文超过60篇，正在编辑出版的著作4部，我们的团队带头人作为主讲嘉宾出席了国内多个大型学术论坛，开展了活跃的对外技术学术交流。我在2013年参加中国建筑学会组织的中国当代建筑师代表团，访问匈牙利、捷克和英国，与当地建筑师协会/学会进行了深入交流，到当今世界顶尖的福斯特事务所、扎哈事务所及罗杰斯事务所交流访问参观，也宣传介绍了我们团队的主要作品。

7. 团队建设：营造人才成长环境

从2004年成立时的6位建筑师，发展到今天的近百人团队，建筑、结构、机电配备完整，2013年又成立ADG建筑创作工作室，ADG·机场院进入了全专业时代。这个事业平台的搭建，凝聚了全体员工的努力与付出。在这个平台里，已经形成了富有层次的团队人才结构，从富有经验的ADG·机场院带头人、副总工、主任、主管、技术骨干到设计人，其中多名团队带头人在行业里已经享有盛誉，可以带领团队在重要项目上夺标、设计及实施。

设计企业最宝贵的资源就是人才，而人才需要培养、需要关心、需要赏识、需要理解、需要尊重，才能够茁壮成长。如何营造良好的人才成长环境，一直是我们思考和关注的重点。我们希望大家共同努力共同进步，长江后浪推前浪，好的传统好的技术要一代一代传承下去。

我们的人才理念是：建立优秀的事业平台，为热爱建筑并立志有所成就的专业人士服务；提供和谐并富于创造力的工作环境；倡导平等质朴的人际关系；注重培训增值、内部晋升，给予员工更大的成长空间；让员工拥有充分的发展机会，在公平环境中尽展才华。

在日益激励的行业竞争当中，建筑专业的前后期分工有其必要性，成立工作室的目的也在于此。然而建筑师的成长必须要经过施工图的训练，所以，需要对此做出可以前后期流动的适当安排。我们鼓励大家在团队中找到最合适自己的岗位，充分发挥自己的特长。

在多元化的业务结构下员工可以开拓视野，积累公建和住宅、大跨度和超高层、大项目和小项目等不同方面的经验。问题的关键是作为企业运作，需要我们进行有效的知识及经验管理，需要对不同业务的特点做好针对性的研究，才能够提供专业化的服务，这也是团队成员必修的功课，否则多元化业务会造成资源浪费、成本提高、服务不到位。

为了营造良好的团队氛围，我们在行政后勤方面不遗余力，想方设法改善工作环境、提供细致的后勤服务，组织内容丰富的业余活动、体育活动、年会活动，组织技术学习、外出参观，专业交流"请进来走出去"。从生活和工作全面关心员工，希望员工真正把 ADG·机场院当成自己的事业之家，与团队共同成长。

8. 未来发展：持续创新之路

从2004年到现在，ADG·机场院已经走过了10个年头，通过每一步扎实的付出，在交通建筑、体育建筑、酒店建筑、商业综合体、住宅地产、总部办公、城市设计、文化建筑、会展建筑等多个领域取得了优异成绩，出色地完成了多项在国内外深具影响的项目设计。

ADG·机场院的作品荣获多项国家、部、省级设计奖，包括国家设计金奖、百年百项杰出土木工程奖、詹天佑大奖、中国建筑钢结构金奖、全国绿色建筑创新奖、全国建筑行业优秀设计一等奖、中国建筑学会建筑创作大奖、广东省优秀设计一等奖等。其中广州亚运馆还荣获AAA 2014亚洲建筑师协会奖、两岸四地建筑设计大奖等国际奖项。

其实，过去的十年仅仅是ADG·机场院历史进程的其中一段，既有十年之前的缘起，更应有十年之后的发展。站在新的起点，我们将以什么心态去迎接未来的挑战？前面的道路也许有很多障碍，总是在十字路口面临选择，人生的经验给我们的启示是"望远路直"，关键是选择什么发展目标。

正如我在本文开头所提到的，展望未来，我们要以中国一流设计企业为发展目标，持守技术本源，继续致力创新，坚定地走好自己的路。在这个快速变化发展的全球化进程中，这必然是一条持续创新之路，在设计理念上创新，在技术科研上创新，在开拓经营上创新，在运营管理上创新，在团队建设上创新，而在创新的道路上将有很多很多的事需要团队去做。未来的ADG·机场院将更加努力，与各界朋友共同推动行业进步，积跬步以至千里！

从事建筑设计行业的建筑师们和工程师们是幸运的，我们的作品可以长久地存留大地。正因为如此，我们的责任也是异常重大的，我们想象中的世界构成了现实世界的一部分，建筑是一个关系到未来世世代代的重大责任，我们确实任重道远！

3

1 香港新福港地产 · 广州鼎峰
SFK · DF Project, Guangzhou

2 罗浮山悦榕庄酒店
Banyan Tree Hotel, Mt. Luofu

3 武汉火车站
Wuhan Railway Station

Within and Beyond a Decade
- Review and Outlook for ADG

Chen Xiong Deputy Chief Architect, The Architectural Design and Research Institute of Guangdong Province (GDADRI)
President, Airport Design Group
Director, ADG Architecture Studio

On Aug 5, 2004 when New Baiyun International Airport was put into use, GDADRI decided to establish the Airport Design Group based on the Airport Design Team, which was called Airport Design Group under GDADRI since 2009. In 2013, ADG Architecture Studio under GDADRI was established based on the creative design team of ADG. Actually, ADG is the abbreviation of Airport Design Group that is originated from Airport Design Team. Of course, ADG could also refer to Architectural Design Group.

On this occasion of the 10th anniversary of official operation of the New Baiyun International Airport and the founding of ADG, it is delightful to see that we have earned some reputation in the industry with our competent design team and a great number of influential projects we have delivered. As leader of our team, I share the pride with our team in what we have accomplished in the past decade. Meanwhile, we will keep working toward our goal to be the first-class design firm in China, and, as earnestly expected by Academician Rong Baisheng, honor our commitment to the technical fundamentals and design innovation.

1. Origin : Role Model and Great Opportunity

As an ideal place for career development, GDADRI fosters and advocates the fine traditions developed by the predecessors. When I started my career in middle 1980's, an era for China to firstly implement its opening up and reform policies, I was lucky to have the opportunity to work with our Chief Architect Guo Yichang and other veteran architects. I was most impressed by Guo's devotion to the architecture creation and design innovation, and his pursuit for his own design philosophy while learning from others as well. Regardless of the project size, he always exerted the same strenuous efforts with the same great passion to devise and deliberate on the design. He was always ready to meet new challenges with his experiences, expertise and wisdom. Guo indeed set up an example for us with his industrious, meticulous, intelligent and modest personality. Later on when I had the opportunity to work with Chief Architect Hu Zhenzhong, Associate Chief Architect Liu Yinpei and other veteran architects, I witnessed again their devotion to the work and appreciate their selfless guidance and support the young architects.

In 1998, the client of Baiyun Airport launched an international design competition for New Baiyun Airport, which attracted many leading international design firms including the consortium of Foster + Partners (UK) + NACO (the Netherlands), ADP, consortium of Parsons (USA) and URS Greiner (USA), and B+H (Canada). GDADRI decided to mobilize a capable design team (the Airport Design Team) to participate in this exciting international competition, which was the first time for GDADRI. We aimed to, by participating in such an important design competition, widen our horizon by learning from the international design firms, meanwhile, showcase the competence and expertise of our design team to win the contract for subsequent design phases. Thanks to the instruction from technical directors of various disciplines and the design team's three months of hard work, we finally presented a design proposal that highly impressed the jury panel and the client. In the autumn of 1998, GDADR was finalized as the local design institute to complete this exciting project in collaboration with an American design firm, a honorable mission for GDADRI to fulfill.

2. First Experience: Terminal 1 of New Baiyun International Airport

For the hard-won Terminal 1 of New Baiyun International Airport, GDADR transferred 80+ capable and experienced architects and engineers from different departments to establish the Airport Design Team that encompassed all disciplines including architecture, structure, plumbing and drainage, electrical, AC, municipal utilities, traffic and budgetary estimate etc. Meanwhile, a technical leading group headed by Academician Rong Baisheng and comprising the chief engineers of all disciplines was also established. From 2000 on, all members of the design team worked under the same root to tackle the various design challenges. Since Aug 2002 after the design was completed, the director of the Airport Design Team

1

1 广州新白云国际机场二号航站楼
Terminal 2, Guangzhou New Baiyun International Airport

2 广州新白云国际机场一号航站楼
Terminal 1, Guangzhou New Baiyun International Airport

2

1

2

and principals of various disciplines all resided in project site to provide design services until the successful one-time airport operation transfer in 2004.

Since 1999 we visited nearly all large airports in USA and the new airports in Asia for researches, including the Hong Kong Airport, Changi Airport, Kuala Lumpur Airport and Incheon Airport. We also visited the Capital International Airport in Beijing and Pudong Airport in Shanghai to obtain the first-hand project information and data. Then we conducted researches for a series of key technologies for the buildings of large span and large space. The technologies we mastered include the Terminal design (terminal composition, passenger/baggage/joint inspection/information flow), the envelop structure technology of large-span building (large metallic roofing system, large-area point-supported glass curtain wall system, tension membrane system), large-span structural technology (truss and lintel-less box-type profile steel panel, 3-pipe shuttle-shaped steel latticed columns), foundation technology for regions with unusually complicated limestone karst development, energy management system (EMS) of large public buildings, AC/fire technology for large and lofty space, as well as specialty technologies relating to the terminal such as the baggage processing, signage system, large-space lighting, counters and public seating etc.

This sino-foreign design collaboration enabled us to accumulate considerable design experiences for large public buildings, in particular in the various disciplines relating to the architectural design of the large space. In addition, we have also obtained abundant experience in management of large project and site services. All these served as solid foundations for us to provide architecture design to complicated large public buildings in terms of technology, management and service. GDADRI thus ranked among the leading design institutes in China.

3. Independent Creative Design: Guangzhou Asian Games Gymnasium

In Oct 2007, the international design competition for Guangzhou Asian Games Gymnasium, the only newly built main venue for Guangzhou Asian Games 2010, was launched. GDADRI passed the prequalification as an independent competitor. Just before the registration for this competition we had talked to two prestigious international firms on design cooperation. Seeking for proper sino-foreign design cooperation has always been our wish, unfortunately no agreement was able to reach due to the disagreement on business terms. After all, we became well experienced with architecture design for large complicated public buildings after our completing the Terminal 1 of Baiyun International Airport, and had confidence in our expertise and competence. So we would not Agree to the sino-foreign cooperation at a low fee, nor just working on the construction document design. Finally the Competition Committee selected 8 international and local design teams to participate in the competition, including 6 international firms or sino-foreign consortiums involving prestigious ones like Zaha Hadid Architects(UK), Arup (UK), gmp International GmbH (Germany) and a French design firm, and two local design firms including us and another Chinese design firm.

It was exciting yet challenging task for ADG-Airport Design Group to participate in the competition on GDADRI's behalf. For a long time, the bid winners of the large complicate public building in China were international design firms and our only chance to win was innovation. The winter at the end of 2007 and the beginning of 2008 was extremely cold with the snow disaster in South China. We met at the office on the 3rd day after the Chinese New Year to work on the 2nd round of design detailing for the competition. We ranked the third place and were shortlisted in the first round, however, were not informed of any subsequent works after submitting the deliverables for the 2nd round. Yet the exciting news came eventually in May 2008 that our design proposal might be selected for implementation. What a nice surprise. Therefore, our independently created design proposal, with its innovative design concept, unique spatial experience and the optimal equilibrium between the identity and feasibility, finally won the competition.

Within a very tight schedule in the following two years, the project design, fabrication and construction were completed to a rather high implementation level, making the project the most eye-catching venue among the others for Guangzhou Asian Games 2010. This has been one of the very few cases where the Chinese architects win an important international design competition with independent creative design proposal which is subsequently realized.

4. All-Win Cooperation: Scientists' Apartment, Guangzhou Science City

In 2006, the international competition for Scientists' Apartment, Guangzhou Science City was launched, for which we constituted a design consortium with Ingerosec. The competition organizer required all registered candidates to submit concepts at the pre-qualification process. Our Japanese partner proposed a site concept where an extra-longish bar building is combined with a spinner-like tower, and included a working model into the brochure. Such unique concept and the way of expression definitely touched the jury panel and we passed the prequalification and became one of the three participants to the competition.

At the subsequent competition stage, we worked closely with our partner who further detailed the site plan and building form, while we took care of the floor plans, apartment type, technical system design and narratives etc. The drafting and typesetting/formatting was decided by incorporating comments from both sides. The model-making company was selected by both partners while we followed up the specific works. Our cooperation during the whole competition period has been pleasant, and the finally submitted design proposal incorporating the concepts and expertise of both parties was selected as the bid-winning proposal. .

Despite of the design adjustment as result of the site change during the project implementation process, the basic elements and features were kept unchanged, i.e. still the bar building + spinner-like tower composition, and the technical design and implementation were based on the new design proposal. Our Japanese partner offered their ideas on façade materiality and both parties worked hard toward perfect detail. The project was finally implemented to a high realization level with high similarity with the one shown in the rendering.

Upon the completion in 2010, the project becomes the landmark in Luogang District. The project featured coated and glazed façade, the balcony comprising the vertical aluminum-alloy profiles and partially steel-constructed roofing. Even without special functionality, complicated technologies and costly materials, the concise and elegant building was well accepted by the Client and all walks of life.

This project was reported and recognized by the professional architectural magazines in both China and Japans including the Architectural Journal (China) and New Architecture (Japan). Due the pleasant cooperation on this project, both parties joined hands again to participate in a number of international competitions and won the international competition of architectural design for Zhuhai Hengqin Development Building and urban design for surrounding area in 2011.

5. Diverse Development: Focus on Both Public Building and Real Estate

The ADG was specially established for the buildings featuring large span and spaces and we actually concentrated on the competition and bidding for airport, railway station, gymnasium and exhibition building. In fact, we participated in the design bidding for more than 10 airports respectively in Kunming, Changsha, Nanning, Chaoshan, Chongqing, Shenzhen, Wuhan, Xiamen, Yantai, Haikou and Pudong of Shanghai. We also participated in several design competitions for Guangzhou Asian Games 2010. We continued our researches on large-span buildings, and expanded our business from the airport terminal to gymnasium, railway station and transportation hub. So far we have completed dozens of influential projects and

1-2 广州亚运馆
Guangzhou Asian Games Gymnasium

3 惠州金山湖游泳跳水馆
Jinshan Lake Swimming and Diving Complex, Huizhou

4 广州科学城科技人员公寓
Scientists' Apartment, Guangzhou Science City

3

4

揭阳潮汕机场航站楼
Jieyang Chaoshan Airport
Terminal and Supporting Works

the representative ones include expansion of Terminal 1 of Baiyun New International Airport, Chaoshan Airport terminal, Guangzhou Asian Games Gymnasium, Huadu Gymnasium for Guangzhou Asian Games, Huizhou Jinshan Lake Swimming and Diving Complex, Wuhan Railway Station (Wuhan-Guangzhou High-Speed Railway) and Ground Transportation Center (GTC) of New Terminal Area, Shenzhen Airport.

We conducted design practices for super high rises as well. Dongguan Hyde Plaza we won in 2005 was our first project developed by private developer. Super high-rise twin towers for hotel and offices were very challenging with their interconnection realized at upper floors. From 2009, we gave more attention to the real estate development as we intended to have more projects and bigger market share, thus more sustaining business. So the real estate developers became potential clients. The design philosophies share many things in common. Our technical experiences and service awareness acquired through the previous public buildings could be referenced for the real estate development projects despite of the difference between the two building typologies. Finally we won a commercial complex at Kuiqi Road, Foshan developed by SFK Properties, a developer from Hong Kong, and another project in Xiankeng Village, Luogang District, Guangzhou of the same developer. We have collaborated with various developers, to name just a few, Kande Hotel Group (Dongguan), New Century (Dongguan), SFK Properties (Hong Kong), Da Heng Qin Company, Poly Real Estate, Evergrand Real Estate, L'sea Real Estate and Banyan Tree Group etc.

Therefore, our multi-facet business structure include both public buildings and real estate development projects, both independent creative work and design collaboration, and both large-span building and high-rises.

6. Emphasis on Technology: Implementation-oriented Academic Research

In addition to a series of important design works that have been completed, we have planned and conducted academic research activities of various forms in view of our business characteristics and by using the platform provided by projects. The activities are thesis writing, monograph writing, patenting, forum organization, review, lecturing, visit and etc. These activities are good opportunities for our employees to better combine the theories into their practices. Moreover, more people have learnt about the key projects completed by ADG.

The papers were started in 2003 when the terminal 1 of New Baiyun International Airport was about to be completed. A paper titled *Terminal of New Baiyun International Airport* was then published in September 2004 on Architectural Journal, a professional architectural magazine of China. And after two years of hard work, the monograph on the same topic was published in 2006. In 2006 and 2008, papers concerning planning and development of New Baiyun International Airport as well as trend and design of airport terminals were published respectively. There were other research achievements on airport terminal design.

In 2010, the Guangzhou Asian Games stadium was completed and won the attention of all sectors of society and major medias. A paper on the design of the stadium was published on *Architectural Journal*. Currently, a book on this project is being edited for publish.

Based on the project of Guangzhou Asian Games Gynasium, a series of research activities were conducted on structural design. Also, many papers on the structural design of the stadium were written and published. The representative ones were *Structural Design of Gymnastic Hall in Guangzhou Asian Games Town*, and *Structural Design of Asian Games History Museum in Guangzhou Asian Games Town* on Journal of Building Structure, and *Researches on Steel Structure Nodes*, and *Application of Cable-stayed Façade in Building Construction* on Building Structure. There invention patents were obtained through researches, including the analysis

1

and computation method for safety of cladding partially applied to composite truss structure, profile steel component that is conductive to reduce stress of the steel component nodes, and reinforced concrete foundation that can improve the aseismic performance.

In the last decade, we have published more than 60 important papers and another 4 works that are being edited for publishing. The leaders of ADG have also attended various large academic forums in China and had academic exchanges with our peers. In 2013, I visited Hungary, Czech and UK as a member of the delegation organized by Architectural Society of China, and had discussions with members of local architectural associations. Also, I visited some top design firms in the world such as Foster + Partners, Zaha Hadid Architects and Richard Rogers Partnership, and introduced some of our major works to them as well.

7. Team Building: Create Environment for Talent Training

Our team has grown from 6 architects in 2004 when it was firstly established to today's nearly 100 architects and structural/MEP engineers. In 2013, ADG Architecture Studio was further established, marking a new era for ADG with fully-equipped disciplines. The formation of this career platform represents the efforts and devotion of all ADG people. So far we have established a complete talent hierarchy including the experienced ADG leaders, associate chief engineers, directors, supervisors, key technicians and designers. Many of the team leaders have well-established reputation in their fields and are capable of leading the team to win, design and realize the important projects.

Talents are the most important assets of a design firm. Talents need to be cultivated, cared, motivated, understood and respected. How to foster a favorable environment for talent cultivation has been the focus of our thoughts and concern. We hope our team could work hard together, grow together and pass our expertise and knowledge to our successors.

When it comes to talents, ADG aims to build up an excellent career platform for professionals who are passionate about architecture and desire to make achievements. It provides a harmonious and creative work environment for its people in which equal and honest interpersonal relationships are encouraged. Training sessions are conducted to improve their skills and facilitate their career development. In ADG, people are given full opportunities to develop and make the best of their abilities.

In the increasingly competitive design industry, it becomes necessary to divide works of architecture design into various smaller stages. The purpose of ADG Architecture Studio just lies in this. However, a good architect must have solid knowledge on construction drawings. Therefore, appropriate arrangements need to be made so that employees could take positions at different stages. Our people are encouraged to find their own suitable position in the team to play their part.

Under the diverse business structure of ADG, our people can broaden their horizon and acquire more experiences from different project typology, such as public buildings vs residential development, large-span structure vs super high-rise, and large project vs small project. The successful operation of a design firm demands effective knowledge and experience management. Furthermore, we need to conduct well-targeted researches in view of the characteristics of different businesses, so as to offer professional services to our clients and avoid the adverse effect of business diversification, such as waste of resources, inflated cost and less satisfactory services.

To foster a positive work atmosphere, we spare no efforts in offering administrative supports to improve our work environment, organize various afterwork activities like sports, annual meeting, training sessions, external visits and technical exchanges with others etc. We hope that our employees will see ADG as their home and grow together with the team.

8. Future Development: Sustainable Innovation

It has been a decade since ADG was founded in 2004. So far we have established a diverse portfolio covering transportation, sports, hospitality, mixed-use development, residences, headquarters office buildings, urban design, cultural buildings and exhibition centers. So far we have completed the designs for dozens of projects that are influential in and outside China.

Over the past decade, ADG has won a series of design awards of national, ministerial and provincial levels, including Gold Medal of National Excellent Engineering Design, 100 Outstanding Civil Engineering Projects from 1900 to 2010, Tien-yow Jeme Civil Engineering Prize, National Building Steel Structure Gold Prize, National Green Building Innovation Award, the First Prize for Excellent Design of National Building Industry, the First Prize of Excellent Design of Guangdong Province etc. In particular, Guangzhou Asian Games Gymnasium was awarded AAA 2014 ARCASIA Award, and Cross-Strait Architectural Design Award.

We believe the past decade is just a small step in ADG's development history. Today when we look back to its establishment 10 years ago, it is more meaningful to think about the way we will take in the future. How should we meet the future challenges? Despite of many difficulties and hard decisions ahead, we are confident that we will fulfill our goal if we have a long-horizon view and establish a development goal first.

As mentioned earlier, it is our goal to become a first-class design firm in China. With our commitment to the technical fundamentals and design innovation, we will work hard toward this goal. To achieve this goal, constant innovation is inevitable in this fast-changing era. The innovations may take place in design concept, technical research, business development, operation management, team building or other aspects, and demand further efforts from our team. So in the future we have to work even harder and join hands with our friends from all walks of life to promote the development of design industry.

We are lucky to be architects and engineers, as our works may last long in the world. In this sense, we shoulder great responsibility as our dreamed world would constitute part of the real world. Architecture concerns many future generations and we still have a long way to go.

2

1 广州花都区东风体育馆
Dongfeng Gymnasium, Huadu District, Guangzhou

2 深圳机场新航站区地面交通中心 (GTC)
Ground Transportation Center (GTC) of New Terminal Area, Shenzhen Airport

好设计 用心做
——《a+a》杂志访谈

1

广东省建筑设计研究院（总院）ADG建筑设计研究院（机场院）凭借出色的设计作品、雄厚的技术实力、优质的专业服务在业界享有盛誉。ADG注重设计原创，注重技术创新，致力绿色环保，以精心设计、至诚服务的团队精神，为业界提供规划、建筑、室内、景观全方位的设计服务，以及可行性研究、钢结构、数值风洞、幕墙、设计监理等高端咨询服务，并具有丰富的中外合作和总包设计经验。ADG设计的工程类型广泛，业务涉及交通建筑、体育建筑、酒店建筑、商业综合体、政府大楼、办公建筑和住宅地产等。

ADG建筑设计研究所（机场院）沿着多元化、专业化、品牌化的道路前行。与客户共同创造价值，为社会进步提供动力。

a+a与广东省建筑设计研究院副总建筑师、ADG（机场院）院长陈雄的对话

a+a：自创办到现在支持ADG的核心价值观是什么?

陈雄：ADG（机场院）是在广东省建筑设计研究院（总院）的一个重要设计团队，2004年开始创办，到现在发展了近10年时间。这个团队的雏形是广州白云国际机场设计团队，即“airport design group”—机场组。而白云机场设计团队的组建是在1998年，到现在整15年。在白云机场竣工之后设计团队的骨干、从各个部门抽调出来的精英及一大批新晋人才逐步构建了ADG。我们的核心价值目标是做好的团队、好的建筑、好的品牌，保持技术的先进性，提供优质的技术服务。

除了北京的T2和浦东的T1，白云机场项目在当时的国内来讲，是比较早的大型综合交通枢纽中外合作设计项目。北京的T2是扩建项目，上海浦东的T1是新建项目，白云机场则是迁建。迁建的难度很高，需要确保一次转场成功。工程建设过程中克服了很多困难和技术上的难题，最终实现了高完成度工程。我们从中获得了很多关于大型工程的经验，也希望将这些经验辐射到其他项目上去。

我们保证高的设计质量的同时，还非常重视项目的附加值。像白云机场航站楼项目，我们开展了一系列的科研活动，这些都是围绕先进技术、设计质量及品牌建设的一个延伸。ADG从最初的机场组到机场所再到机场院，最终成为国内第一个以机场团队命名的设计院。在民航业内ADG代

1 揭阳潮汕机场航站楼
Jieyang Chaoshan Airport Terminal and Supporting Works

2 广州亚运馆
Guangzhou Asian Games Gymnasium

3 广州新白云国际机场二号航站楼
Terminal 2, Guangzhou New Baiyun International Airport

表了机场院，英文是airport design group，而其之外ADG的含义就是：Architecture Design Group，这样更有利于业务的持续开展以及及与市场的接轨。我们有自己一批建筑方面的人才，首先更加强调原创设计，同时也有比较稳定的高水平国际合作伙伴。

a+a：ADG主要以设计大型民用建筑为主，这种大型工程从设计到完工整个项目团队是如何运作的呢?

陈雄：ADG所在的广东省建筑设计研究院是成立了60多年的国内大型设计院，除了建筑专业以外，结构专业和机电专业技术能力也很强大。大型项目功能复杂，涉及专业比较多，这样正好发挥了大型设计院的优势。我们总院的专业，除了传统的建筑、结构、机电、建筑经济，还有市政、交通、园林景观、勘察等等这些。

大型工程从设计到完工整个项目团队是如何运作的呢？以白云机场2号航站楼为例，除了集合了ADG的优势资源之外，还从院里面其他专业整合了大约160人，作为一个团队全部在ADG办公区内进行集中的管理。就是以项目为单元进行管理，根据项目的规模决定项目团队的规模。项目运行的模式除了这样的专组之外，还有把关的团队，同时在异地的同事也可以通过网络进行联合作业。

a+a：设计建造大型民用建筑与一般的建筑除了规模之外，还有什么区别?

陈雄：主要的区别是功能及流程方面的思考要更有深度和广度，其系统和设备繁多，技术和材料新颖，往往涉及很多投资，因此对城市和公众的影响力更大。这也从另外一侧面说明建筑师的社会责任更加重。大型项目以机场航站楼为例，它和人们生活的方方面面都有关系。虽然它是交通建筑，但还有像计时旅馆的居住、大型免税购物等其他方面的需求。在这里的使用者有旅客、航空公司、机场业主、联检单位（检疫、边检、海关）等，他们的需求都要满足。同时，这些大型项目的建设周期安排都比较赶，需要我们突破时间的制约，高效率地工作。大型民用建筑投资都很大，在审批的概算内实现建筑设计，这体现了设计机构的综合能力。而它们的绿色环保要求更高，降低能耗的实际意义非常突出，还是以2号航站楼为例，它的定位是中国绿色三星，确定目标后，根据国家的规范进行各方面的分解，达到更好的节能效果。当然，大型民用建筑会聚集很多的先进技术和新的材料，对于技术进步产生更大的引领作用。

a+a：怎样加强企业文化的培养呢?

陈雄：ADG的积累和传承、对设计品质的追求、对技术先进性的追求是基于整个广东省建筑设计研究院多年的积淀。文化的培养在于历史的传承，同时我们也关注建筑设计的前沿发展以及市场需求的更新，做到既有传承也有创新。而且我们为员工提供大量开阔视野的职业机会，对员工成长的很多关怀，注重对他们的培训，创造好的发展空间。我们构建了一些专业技能体系，员工在日常的项目设计中遇到一些困难，或者需要一些经验指导，可以从体系中得到答案，这样他们工作起来就会更加顺利和舒心，也有利于提高业界对团队能力认可。我们还比较关心员工的业余生活，安排了很多体育活动。

a+a：荣誉对于ADG代表什么?

陈雄：首先要精心做好每一个设计，开始做项目的时候，我们没有想到一定得什么奖励。好的项目需要天时、地利、人和，荣誉只是代表对过去我们一贯的坚持和努力的肯定。我还是强调建筑师的责任重大，他们想象的世界将会通过作品构成现实世界的一部分，进而影响到社会生活的各方面。对于ADG来说荣誉是对团队工作和所负社会责任的认可，是一种激励，尤其是对年轻同事。

a+a：ADG的未来是怎样规划的?

陈雄：主要有五个方面，一是加强原创能力的培养，进一步提高建筑的原创能力，同时我们国家也需要更多的原创作品；二是继续关注建筑创作前沿的发展，包括国外的一些先进设计理念和技术，把握好建筑设计的潮流方向；三是在业务类型多元化的基础上，树立好自己的品牌形象，使业务发展更加稳健；四是长远规划还包括对员工发展的持续关注，设计企业人才是特别重要的，因此员工成长需要密切的关注；五是提高项目运营效率，一如既往地保证项目设计质量。最终还是回到了所有关于设计团队核心理念的原点，就是要做好的品牌，好的作品。

2

3

Design with Heart and Soul

- a+a interview

ADG Architectural Design & Research Institute (Airport Design Group, "ADG") of Architectural Design & Research Institute of Guangdong Province (Headquarters) (GDADRI) has established its reputation in the industry with the impressive works, expertise and professional services. Collaborating extensively with international design firms and working as general contractor for design, ADG offers full spectrum of design services covering planning, architecture, interior, landscape, etc., as well as consultancy services related to feasibility research, steel structure, numerical wind tunnel simulation, curtain wall, design supervision, etc. ADG values originality and technical innovation in design and implements the green and environmentally-friendly approahces. With concerted team spirit and commitment to perfection and excellence, ADG offers designs for a great variety of projects, including, transportation buildings, sports venues, hospitality, mixed-use development, government buildings, office buildings and residential development, etc.

Guided by a more diverse, specilized and brand-centered strategy in future developmment, ADG will join hands with the clients to create value and drive the social progress.

a+a's interview with Chen Xiong, Deputy Chief Architect of GDADRI and President of ADG (Airport Design Group)

a+a What is the core value that has been propping up ADG since its establishment?

Chen: It is nearly ten years since its incorporation in 2004 and now it has become one of the important design teams of GDADRI (Headquarters). The predecessor of the team is Guangzhou Baiyun International Airport design team, namely, "airport design group"- airport group, which was set up in 1998, more than 15 years ago. After the completion of the airport, the backbone of the design team, together with the elite selected from different departments and a bunch of talented young people, established ADG step by step. Our core value and goal is to be a great team that designs great buildings to build great brand and maintains its edge in technologies to provide best technical services.

The Baiyun Airport Project was another earliest sinno-foreign collaboration on large-scale comprehensive transportation hub in China, other than T2 in Beijing and T1 in Pudong Shanghai. T2 was an expansion project and T1 a new project, whereas Baiyun Airport was a relocation project of exceptional difficulty which must be guaranteed successful in one stroke. During the project construction, we overcame many difficulties and finally realized the porject to a high level. From this project we acquired a lot of experiences on large projects that serve as reference for other projects.

In addition to high design quality, we emphasize adding value to the projects. Take the Baiyun Airport Project as example. We conducted a series of scientific researches which revolved around the advanced technology, design quality and branding building. ADG developed from the airport group to the airport office and then the Airport Design Group and finally became the first domestic design institute named after the airport team. In the line of civil aviation, ADG represents the Airport Design Group. ADG is the acronym for both airport design group and Architecture Design Group, which will facilitate the continuous business development and alignment with the market. With a flock of architectural professionals, we put a premium on originality in design and have developed fairly stable cooperative bonds with some high-class international partners.

a+a ADG focuses on designing large-scale civil buildings. How does the project team run a large-scale project from design through completion?

Chen: ADG is affiliated to GDADRI which is a large design institute established more than 60 years ago and has strong technical capacity not only in architecture but also in structure and electro-mechanics. Large-scale projects usually have complicated functions and involve a host of specialties, which gives full play to the strengths of large institutes. Apart from the typical specialties of architecture, structure, electrical, mechnical and architectural economics, our headquarters have other specialties such as municipal utilities, transportation, landscape and engineering investigation, etc.

As regards how the project team runs a large-scale project from design

1

1 惠州金山湖游泳跳水馆
Jinshan Lake Swimming and Diving Complex, Huizhou

2 珠海横琴发展大厦
Zhuhai Hengqin Development Building

3 广州科学城科技人员公寓
Scientists' Apartment, Guangzhou Science City

2

3

through completion, let's take the Terminal 2 of the Baiyun Airport for example. For this project, we gathered the best resources of ADG together with around 160 professionals of different specialties from GDADRI to form a team and had them centralized in the office area of ADG. That is, the teams are managed by projects and the scale of a project determines the scale of the project team. Besides a project-based team, the project operation mode also contains a control team and collaboration with colleagues at other locations via the internet.

a+a Other than scale, what are the differences in design between large-scale civil buildings and ordinary buildings?

Chen: The large-scale civil buildings usually demand deeper and broader deliberation over functions and processes, and involve sophisticated systems/equipment, new technology /materials, and huge investments. Therefore, the projects are more influential and the architects have to shoulder more socail responsibilities. For example, the airport terminal as a large project, it is related to all aspects of the people's life. Though designed as a transportation building, it also accommodates short-time hotel, duty-free shops, and other functions. The users of this building include passengers, airlines, owner of the airport, joint inspection authorities (quarantine inspection, border inspection, customs), etc., and all their demands must be met. Moreover, the construction schedules of such projects are usually rather tight, which requires us to work at fast pace. Such projects also involve huge investment and the architectural designs must be realizable within the given budget. In addition, their requirements on green technology and environmental protection are particularly high, so energy efficiency and reducito n are of great practical significance here. For example, the Terminal II of Baiyun Airport aimed to be a 3-star green building. Under this target, the relevant Chinese codes were analyzed and applied to different aspects of the design for better energy efficient result. Of course, as large-scale projects pool a number of cutting-edge technologies and new materials, they can play a more prominent leading role in the development of technologies.

a+a How do you cultivate the corporate culture?

Chen: ADG's project experiences and pursuit for design quality and up-to-date technologies are based on the legacy of GDADRI. Our culture is originated from both legacy and innovation. We offer extensive career opportunities for our employees to broaden their horizon. We set up some systems of specialized techniques so that when our people have questions or need guidance about the design, they can find answers in these systems, which makes their work easier. We also organize many sports activities for our staff so they can relax and enjoy themselves after work.

a+a What does an award mean for ADG?

Chen: First of all, it means we have to work meticulously on each project. When we start a project, we don't think of winning any award. It requires good timing, situation and people to make a good project, whereas an award only means the recognition to our unremitting perseverance and effort. I have to underscore again the significant responsibility of architects, in that the world devised by them will become part of the real world through their works and then influence all aspects of the social life. For ADG, an award means the recognition to our work and social responsibilities,and will encourage us to do a better job in future, especially for our young people.

a+a How does ADG plan for its future?

Chen: we will focus on five aspects in future. The first is to enhance our original creative team, as our country is in want of more original design works; the second is to keep a close eye on the latest developments in architectural creation, including the advanced design concepts and techniques from overseas, to grasp the trend and direction of architectural design; the third is to establish our own brand name through business diversification to ensure steady business development; the fourth is to make a long-term plan for our career development of our staff, in that people are of exceptional importance to a design firm; the fifth is to enhance project operation efficiency and ensure project design quality as always. So finally we get back to the origin, i.e., the core tenet of the design team, that is, to build great brand and design great projects.

Selected Works　作品集萃

Selected Works 作品集萃

广州新白云国际机场一号航站楼
Terminal 1, Guangzhou New Baiyun International Airport

1998-2004　广州 花都 / Huadu Guangzhou

夜幕降临，随着一架架飞机降落在新机场，随着老机场跑道灯光的渐渐熄灭，一个令世人瞩目的时刻到来了：2004年8月5日，广州新白云国际机场正式启用。作为我国三大枢纽机场之一，一个按照中枢理念设计、建设、营运的崭新机场，经过近10年的筹划，4年的规划设计，3年零10个月的建设，终于在花都一片宽阔的土地上崛起。航站楼造型新颖，具有明显的中轴线，展现了强烈的标志性。在蓝天白云之下，地平线上一组流畅有力的弧线，勾勒出独特的建筑形象，高低起伏，充满动感，在亚热带植物的衬托下体现出中国南大门的雄伟气势。新机场既是广东建筑业的一大成就，也是我国民航史上一个重要的里程碑，反映了中国特别是广州最新的建筑技术水平，展示了新世纪中国大型标志性建筑的先进性与独特性。

航站楼位于间距为2200m的东、西跑道之间，设计年旅客吞吐量2500万人次。航站区中央留出用地兴建航管楼、塔台、停车楼和机场酒店。航站楼构型在国内外机场中独一无二，采用"主楼+双向连接楼+指廊式"构型。航站楼由主楼、东连接楼、西连接楼和四条指廊组成，拥有总计66个机位（近机位46个，远机位20个），高比例的近机位方便旅客。其最大特点是出港旅客在主楼三层办理登机手续，在指廊分散候机和登机，到港旅客由指廊分别经东、西连接楼首层提取行李离开。主楼具有可从南、北方向双向进入的特点，具有双倍的车道边。航站楼扩建时向北发展，不影响首期营运。这是国内第一个将地铁站设在出港大厅下面的航站楼，毫无疑问是机场和地铁紧密结合的杰出工程范例。

本项目在中国首次大面积使用预应力自平衡索结构点支式玻璃幕墙，是迄今为止世界上该类型幕墙面积最大的单体工程。张拉膜用于覆盖复杂的曲面造型，也是在中国机场中第一次使用。这还是中国目前在岩溶地区兴建的最大规模的民用建筑项目，在中国首次使用三管梭形钢格构人字柱，以及大跨度屋面无檩式箱形压型钢板。

航站楼自然光线充足，几乎所有的公共空间白天都无需人工照明。连接楼是东、西向建筑，在弧形玻璃幕墙的顶端特别设计了光控可调电动遮阳百页。设计生态、环保、节能，环境舒适。新白云机场获得中国首届绿色建筑创新奖。

On August 5 2004, Guangzhou New Baiyun International Airport was officially put into use. As one of the three hub airports in China designed, built and operated following the hub concept, it finally rose in Huadu District after years of preparation, planning and construction. The new airport terminal is a highly iconic building with a novel shape, a clearly defined central axis, and an undulating and dynamic building image portrayed by a series of sweeping arc lines stretching above the horizon. It is not only a great achievement in Guangdong's construction industry but also a significant milestone in China's aeroplane history. It represents the latest architectural standards in China, particularly in Guangzhou, and showcases the high performance and uniqueness of large iconic buildings in China in the 21st century.

With a designed annual throughput of 25 million person-times, the terminal building rises between the east and the west runway which are 2,200m from each other. The ATC building, tower, garage structure and airport hotel are planned in the middle of the terminal area. The terminal in a unique composition, i.e. Main Building + East/West Connecting Buildings + Four Piers has totally 66 stands (46 near stands and 20 remote stands). The high percentage of near stands offers convenience to passengers. The Main Building is accessible from both the north and the south hence offers double arrival and departure curbsides. The future northward expansion will not impact the Phase I operation. As the first terminal in China which has a metro station below the departure concourse, it sets up a noticeable example for integration of airport with metro.

As the first project in China to use pre-stressed self-balanced cable and point-supported glass curtain wall for the façade, the terminal building boasts the largest facade area among the singular projects of the same kind in the world. It is also the largest civil building ever constructed in karst area by now, and the first project in China that uses the herringbone three-tube fusiform composite column and large-span purlin-free roof box profiled steel sheet.

Almost all public spaces in the Terminal can enjoy natural light. Light-operated adjustable electric shading louvers are especially installed on the top of the arc glazed curtain wall. The project has won the first "National Green Building Innovation Award".

项目地点：广州市北部，白云区人和镇与花都区新华镇的交界处
设计时间：1998—2002年
建设时间：2000—2004年
用地面积：922,122m²
建筑面积：353,042m²
建筑层数：地上3层，地下2层
建筑高度：55.88m
合作单位：美国PARSONS公司 + URS Greiner公司
曾获奖项：2011年评为中国"百年百项杰出土木工程"
2010年中国建筑学会建筑创作大奖
2007年度广东省优秀工程技术创新奖
2006年度"全国优秀工程勘察设计金质奖"
2005年获得第五届"詹天佑土木工程大奖"
2005年"全国十大建设科技成就"称号
2005年获得"首届全国绿色建筑创新奖"
2005年获得广东省优秀工程设计一等奖

Location: North of Guangzhou, at the junction between Renhe Town, Baiyun District and Xinhua Town, Huadu District
Design: 1998-2002
Construction: 2000-2004
Site: 922,122m²
GFA: 353,042 m²
Number of floors: 3 above-grade floors and 2 below-grade floors
Building height: 55.88m
Partner:PARSONS + URS Greiner (USA)
Awards:
"100 Outstanding Civil Engineering Projects from 1900 to 2010" (2011)
ASC Architectural Creation Award (2010)
Excellent Engineering Technology Innovation Award of Guangdong Province (2007)
Golden Prize for National Excellent Engineering Exploration and Design Award (2006)
The 5th Tien-Yow Jeme Civil Engineering Prize (2005)
Top 10 National Construction Technology Achievement Award (2005)
The First "National Green Building Innovation Award" (2005)
The First Prize for Excellent Engineering Design Award of Guangdong Province (2005)

1　白云机场总平面图（含一号、二号航站楼）
Master Plan of Baiyun International Airport (including Terminal 1 and 2)

2　主楼出发大厅中央的张拉膜采光带引导旅客分别前往东/西连接楼
The central daylight strip made of tension membrane in departure hall lead passengers to the east/west connecting building

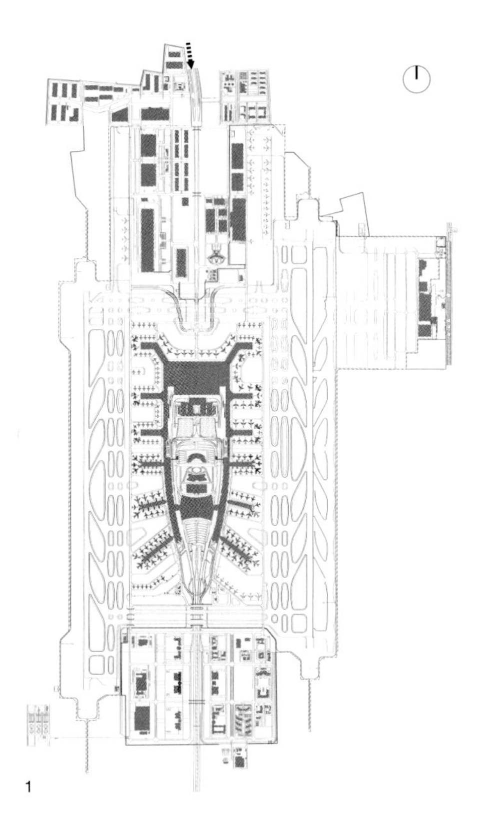

1

2

绚丽的航站楼夜景：主楼大面积点式玻璃幕墙及张拉膜雨篷
Magnificent terminal night view: large-area point-supported class curtain wall and tensile membrane canopy of main building

8
7

1

2

3

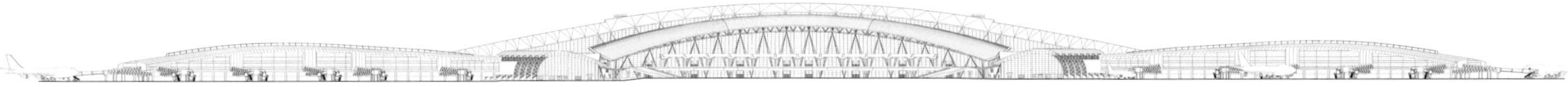

6

4

5

1 “主楼+双向连接楼+指廊式”的独特构型
Unique Configuration of "Main Building + Two-way Connecting Building + Pier"

2 晶莹剔透的航站楼
Crystalline terminal

3 一组流畅有力的弧线构成航站楼独特的形象
Smooth and overwhelming arc lines portray a unique image of terminal

4 航站楼空侧
Airside of Terminal Building

5 东西连接楼夜景：“老虎窗”及张拉膜雨篷
Night view of east-west connecting building: "Dormer" and tensile membrane canopy

6-7 立面图
Elevation

7

1 自然光下的候机厅
Daylit Waiting hall

2 办票大厅与带广告的办票岛设计
Check-in hall and check-in counter with advertisement

3 出发大厅夜景：以间接照明实现舒适的空间氛围
Night view of departure hall: create comfortable spatial atmosphere through indirect lighting

4 主楼剖面图
Section of the main building

5 候机厅的张拉膜采光带：指引旅客流程并避免眩光
Tensile membrane daylighting strip in waiting hall: guide the passenger circulation and avoid glare

1

2

3

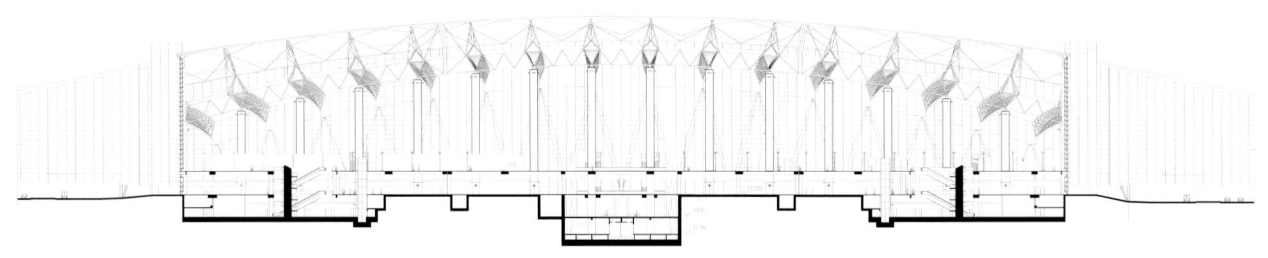

4

B123
HU6001T 18:00
海口/Haikou
B125-131
B125
B126
B124

1

2

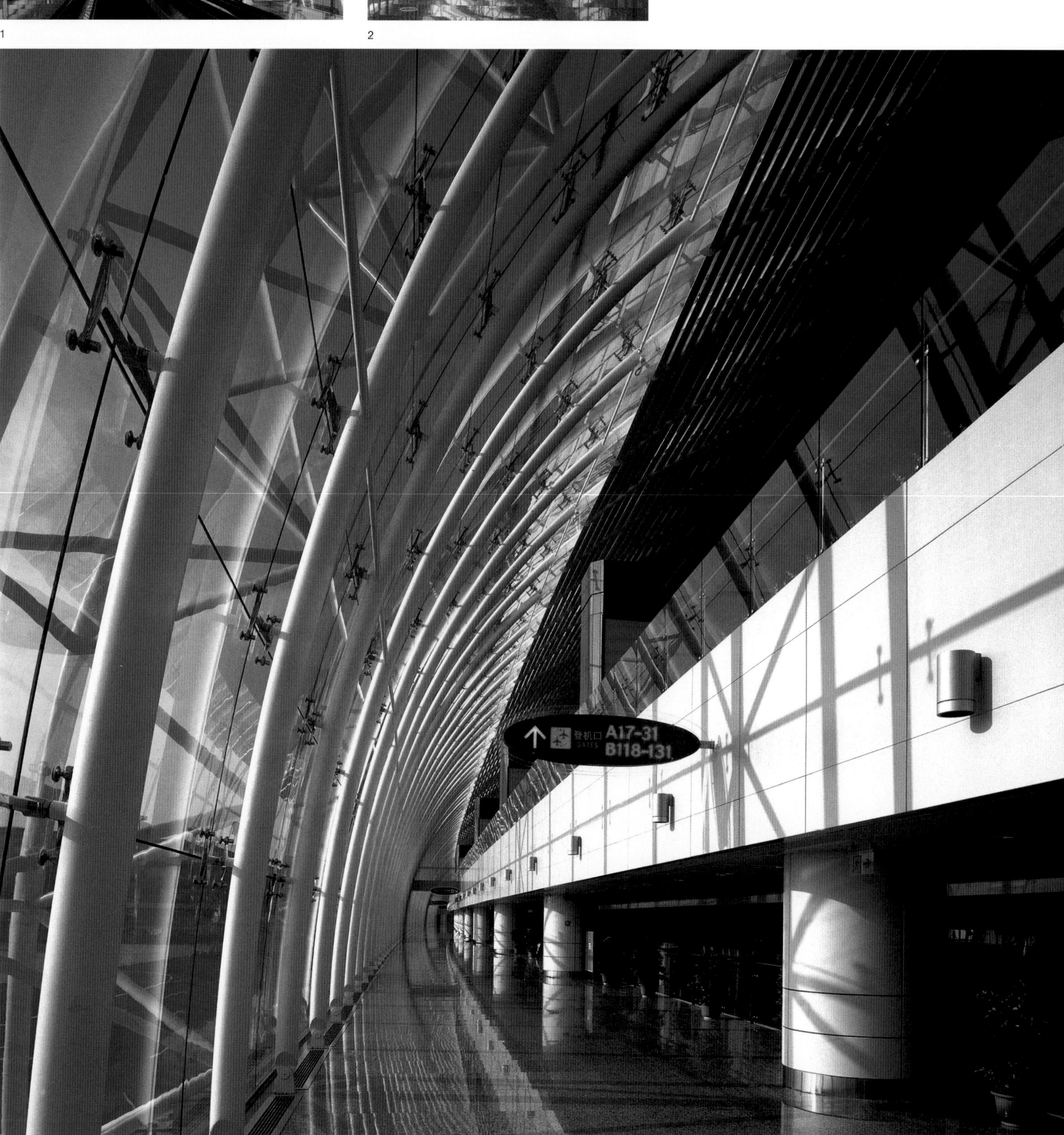

3

4

5

1 连接楼三层出发过厅：弧形点式玻璃幕墙及电动遮阳百叶
Departure foyer on F3 of the Connecting Building: the curve point-supported glass curtain wall with electric sun-shade louvers

2 高大舒适、自然采光充足的迎客大厅
Lofty, comfortable and light-flooded arrival hall

3-4 连接楼二层到达走廊：弧形点式被动幕墙使空侧机坪景观开阔舒展
Arrival corridor on F2 of the Connecting Building: curve point-supported glass curtain wall allows for the open and far view to the airside parking apron

5 主楼预应力自平衡索结构点式玻璃幕墙
Main building features the point-supported glass curtain wall with pre-stressing self-balancing cable net

夜色下“水晶宫”般的航站楼美轮美奂
A magnificent night view of terminal

广州新白云国际机场东三、西三指廊及相关连接楼
East III, West III Pier and Connecting Building, Guangzhou New Baiyun International Airport

2005-2009　广州 花都 / Huadu Guangzhou

2004年8月新白云国际机场落成启用，为广州航空事业带来了新一轮的发展高潮。2005年旅客量达2339万人次，比2004年的2032万人次增长了14%。如此迅猛的发展势头，可以预见2006年白云机场的年旅客量将突破一期航站楼的设计容量——2500万人次。因此，航站楼的扩建迫在眉睫。根据发展预测和总体规划，本期扩建工程沿东、西连接楼向北发展，各增加一条国内指廊，增加20个近机位，增加候机厅及行李提取盘，而办票和行李分拣设施则利用一期主楼预留的发展空间。一期工程加上扩建工程构成完整的T1航站楼，运输能力提高至年旅客吞吐量3500万人次，满足广州亚运会的航空需求。

为实现一期航站楼和扩建工程的整体性，采用了"延续"的建筑语汇，同时又对室内空间元素进行了优化整合，在一期工程的基础上有所"超越"。保持空侧、陆侧容量的平衡，做到功能完善、流程顺畅，提高服务水平，是本次工程的出发点及重点。

连接楼老虎窗结构与主体钢结构合二为一，不需设横向支撑，结构更合理，迎客厅的空间也更为纯净。三层楼板出挑加大获得更多的商业面积，而增加迎客厅进深则使空间更加舒展。还加大了一层的商业面积，在现有迎客厅与扩建的迎客厅之间设置了一个商业中庭，中庭内种植绿化，既改善商业环境，又吸引迎客厅的旅客到此消费。

为节省空侧机坪面积，增加近机位，在指廊三层增设了室内坡道，缩短了登机桥固定端的长度。在候机厅结合室内弧形屋面的特点，在空间最高处设计了局部的四层作为头等舱和商务舱旅客休息室。指廊三层候机大厅宽度比一期指廊增加了11m，且柱跨增加至18m，玻璃幕墙形式也更为通透，使整个空间更加宽敞明亮。指廊端头登机口主要服务A380和B747等大型飞机，因此，在指廊端头布置了较大范围的座椅区。

结构设计首次采用了大跨度预应力空心钢管桁架及铸钢节点等新技术，玻璃幕墙采用了单索预应力索网结构等新技术。

设计了地下设备管沟，各设备专业的主要管线水平布置，再垂直上行至各楼层，节省吊顶空间并保证了到达层净高，也有利于运营管理和可持续发展。

扩建后的T1航站楼为广州亚运会提供了优质空港服务。

In August 2004, Guangzhou New Baiyun International Airport was completed and put into use, making a new peak of Guangzhou's aviation business. The total passenger throughput in 2005 reached 23.39 million person-times, an increase of 14% compared to the annual throughput of 20.32 million person-times in 2004.

To ensure the integration between Terminal 1 and the expansion part, the architectural vocabulary of Phase I has been continued while optimizing and refining the interior spatial elements. The key of the project is to balance the capacity on the landside and the airside, meanwhile, realize the well-established functions, smooth work flow and higher service quality.

The dormers and main steel structure are integrated to avoid lateral support, thus a more rational structure and pure space in the meter-and-greeter hall. The cantilevered floor on F3 offers more retail area and creates more comfortable space for the meter-and-greeter hall. The retail space on F1 is also increased. A commercial atrium is provided between the existing and proposed arrival halls, where plants will be grown to better the environment and attract the passengers from the arrival hall to shop here.

Interior ramps are provided on F3 of the pier and the fixed end of the boarding bridges are shortened to spare the space in the airside apron to accommodate more near stands. Partial F4 is created at the highest part as the first-class and business-class passenger lounge. Waiting hall on F3 is 11m wider than that of Phase I, with the column grid increased accordingly to 18m.

Pipe trenches are designed underground, save ceiling space to ensure a sufficient height clearance for the arrival hall and facilitate the operation management and sustainable development. The expanded T1 terminal, upon completion, has offered quality service for Guangzhou Asian Games.

项目地点：广州市北部，白云区人和镇与花都区新华镇的交界处
设计时间：2005—2008年
建设时间：2005—2009年
建筑面积：148041m²
建筑层数：地上三层（指廊局部四层），地下一层
设计顾问：美国YM公司
曾获奖项：2011年广东省优秀工程设计二等奖

Location: North of Guangzhou, at the junction between Renhe Town, Baiyun District and Xinhua Town, Huadu District
Design: 2005-2008
Construction: 2005-2009
GFA: 148,041m²
Number of floors: 3 above-grade floors (partially 4 floors in Pier) and 1 below-grade floor
Design Consultant: YM (USA)
Awards: The Second Prize for Excellent Engineering Design Award of Guangdong Province (2011)

1 扩建指廊更为宽敞舒适，与商业密切结合。张拉膜采光带是保留的设计元素
The expanded concourses are more spacious and comfortable with closer connection with retails. The reserved design elements include the tensile membrane daylighting strips

2 连接楼二层到达走廊仍采用弧形点式玻璃幕墙，另配电动遮阳帘
The arrival corridor on F2 of the Connecting Building still uses the curve point-supported glass curtain wall with the additional electric sun-shade strip

1

2

2

3

1

1 连接楼仍保留"老虎窗"及张拉膜雨篷
The Connecting Building still features "Dormer" and tensile membrane canopy

2 连接楼三层出发过厅
Departure foyer on F3 of the Connecting Building

3 迎客大厅结构更加整齐简洁
The Arrival Hall is neater and better organized

广州新白云国际机场二号航站楼
Terminal 2, Guangzhou New Baiyun International Airport

2005— 广州 花都 / Huadu Guangzhou

广州亚运会以来新白云机场继续快速发展，在2013年旅客量超过了5000万人次，已经达到1998年规划的总量，进入世界机场前15名。因此，T2航站楼及配套设施的建设势在必行。这是目前我国在建的规模最大的航站楼，面积超过63万m²，配套的交通中心（GTC）及停车楼达21万m²，设计容量为年旅客量4500万人次，近机位70个，新白云机场总计年旅客量将达到8500万人次。目标是打造亚太地区连接各大洲新的航空枢纽，朝着世界级机场迈进。T2航站楼于2013年动工，计划2018年完成施工投入使用。

T2航站楼在T1航站楼北面进行扩建。设计团队从2005年开始T2航站楼规划设计研究，保持南北贯通的航站区布局，北进场车流通过主楼下面的隧道与南侧的陆侧交通路网连接，经过高架路组织分流进入航站楼、交通中心（GTC）或停车场。航站楼规划将"分离站坪"改为"北站坪"概念，采用"指廊式＋前列式"混合构型，获得多个紧靠主楼北站坪的前列式大型机位，国际－国内可转换使用，提供使用的灵活性并有效缩短了旅客步行距离。T2航站楼包括主楼、六条指廊及北指廊，其中东四和西四指廊分期建设，近机位比例维持新白云机场一贯的高水平。旅客流程与T1航站楼完全不同，出发旅客从南侧高架桥车道边进入主楼办票大厅，往前进入安检大厅或联检大厅，经过检查后再分流到达候机厅。到达旅客集中在主楼提取行李，国际和国内迎客厅均在主楼首层。出发和到达旅客流程简洁方向清晰。在本期建设中，预留了旅客捷运系统（APM）的空间及结构荷载，为将来出发、到达特别是中转旅客提供高水平服务。在主楼南侧的交通中心（GTC）实现了大巴、出租车、社会车、地铁和城轨多种交通方式的快速换乘，达到完全人车分流。

T2航站楼的设计目标是拥有流畅的旅客流程和完善的功能设施；与T1航站楼和谐一致的建筑造型，保留了弧线形的主楼和人字形柱及张拉膜雨篷这些特有元素；体现岭南地域特色的花园空间及装修设计；强化商业资源为提高非航收入奠定基础；增加设计弹性应对未来需求变化；注重绿色环保节能设计，达到中国绿色建筑三星标准；成为展示公共艺术的枢纽门户；反映当今中国最新的建筑技术水平。

Since the Guangzhou Asian Games, Guangzhou New Baiyun International Airport has experienced a rapid development. In 2013, its passenger throughput exceeded 50 million person-times which was the peak capacity projected in 1998 plan. Therefore, it has become imperative to construct Terminal 2 and its supporting facilities. Terminal 2 is currently the largest terminal under construction in China, which has a GFA of over 630,000 m² and a supporting ground transportation center (GTC) including a garage structure of 210,000m². It is designed with an annual passenger throughput of 45 million person-times and 70 near stands. Upon completion of Terminal 2, the total annual passenger throughput of Guangzhou New Baiyun International Airport will reach 85 million person-times. The construction of Terminal 2 has commenced in 2013 and is expected to complete in 2018 when it can be put into use.

Terminal 2 is located to the north of Terminal 1. The design team started planning and design of Terminal 2 since 2005, aiming to maintain a terminal area layout that runs in a north-south direction. In planning of Terminal 2, the original concept of "a separated apron" is replaced by the concept of "a north apron" and a mixed "Pier + Linear Composition" is adopted, so as to have multiple large linear stands that are close to the north apron of Main Building. The stands can be used for either international or domestic flights, enabling more flexible operation and effectually shortening the walking distance of passengers. Terminal 2 consists of Main Building, 6 Piers and a North Pier. Passenger circulations of Terminal 2 are totally different from those of Terminal 1. Spaces and structural loads are reserved for automated people mover system (APM) during construction in current phase, which will provide quality service to departing and arriving passengers and particularly transit passengers in the future. The GTC located to the south of Main Building allows for rapid interchange of people among multiple transportation modes such as bus, taxi, private car, metro and track traffic, so as to completely separate pedestrian circulations from vehicular circulations.

The design objectives of Terminal 2 are to create smooth passenger circulations and fully functional facilities; provide a building shape that is in harmony with Terminal 1; reflect the local features of Lingnan; increase non-aviation-related revenues; achieve China's 3-star Green Building standard; create a gateway for display of public artworks; and reflect the highest building technology level of China.

项目地点：广州市北部，广州新白云机场航站区
设计时间：2005—2014年
建设时间：2013年至今
建筑面积：63.4万m²（2020年，不含东四西四指廊）
建筑层数：地上5层，地下2层（GTC）
建筑高度：44.675m
顾问机构：美国MA公司、美国L&B公司等

Location: North of Guangzhou, at the Terminal Zone of Guangzhou New Baiyun International Airport
Design: 2005-2014
Construction: 2013 to date
GFA: 634,000m2 (2020, excluding East IV, West IV Pier and relevant Connecting Building)
Number of floors: 4 above-grade floors and 2 below-grade floors (GTC)
Building height: 44.675m
Consultant: MA, L&B (USA), etc.

1

2

1 T2航站楼采用“指廊式+前列式”混合构型
Terminal 2 features a mixed “Pier + Linear Composition”

2 T1与T2航站楼作为一个整体设计并实施
Terminal 1 and 2 are designed and implemented as a whole

1

2

3

1 T2航站楼造型更为简洁流畅，保留了弧形屋面、人字形柱及张拉膜雨篷的设计元素
Terminal 2 features more concise shape with elements like curved roof, herringbone columns and tensile membrane canopy

2 航站楼与交通中心连接处效果图
Rendering of connection between the terminal and the transportation center

3 陆侧出发层车道边近景效果图
Close view rendering of the driveway side on the landside departure floor

1

2

3

4

1 具有波浪形吊顶的办票大厅，空间设计简洁流畅，色调明快，外部绿化景观内渗
Check-in hall features wavy ceiling, flowing space and lively tone, with exterior greenery gently oozing into the interior

2 国内候机大厅：候机厅内设有商业，等候空间与商业空间紧密结合，丰富人的活动
Domestic waiting hall is provided with retail space which is closely combined with the waiting space to offer more activities to passengers

3 国际出发商业空间设有天窗、侧窗，引入花园景色及阳光，使传统的商业空间焕然一新
Retail space in international departure hall presents a refreshing look with the garden view and sunlight brought into the interior via skylights and side windows

4 国际到达走廊采用大面积落地玻璃幕墙，国际旅客到达广州后第一眼就能看到走廊外特具岭南特色的园林景观，充分体现新白云国际机场作为广州门户的作用
The international arrival corridor features ceiling-to-floor glass curtain walls, impressing the international passengers with attractive Lingnan-style garden views to the international passengers upon their arrival and embodying the airport's role as the city's gateway

揭阳潮汕机场航站楼及配套工程
Jieyang Chaoshan Airport Terminal and Supporting Works

2007-2011　揭阳 炮台镇/ Paotai town Jieyang

揭阳潮汕机场是省内继广州新白云机场和深圳宝安机场之后的第三大机场，主要服务粤东地区及闽南部分地区。潮汕机场的落成，使该区域摆脱军民合用机场的限制，真正拥有一座现代化的民用干线机场，对于形成粤东地区交通一体化和经济一体化，推动粤东地区经济发展具有十分重要的意义。

潮汕机场航站楼属中型机场规模，将分三期建设。一期航站楼采用了"指廊式+前列式"的复合构型，在相同用地条件下提供最多的近机位，用地效率高。三条与主楼连接的指廊长度约200m，步行距离适中，旅客流程便捷清晰。将来在不停航的前提下，主楼和指廊都可以按需要进行扩建。扩建后仍然是一座具有完整性、一体化的航站楼。航站楼内部空间设计借鉴了传统民居的庭院概念，在主楼和指廊之间设有一个悬空式花园。无论在办票厅、候机厅还是到达走廊、行李提取厅，旅客都可以欣赏到美丽的花园景色，给旅客留下深刻印象。同时，花园还促进航站楼的通风，结合大挑檐、室外遮阳、局部天窗等建筑元素，降低建筑能耗。

潮汕机场于2011年12月建成使用，为来往潮汕地区的旅客提供了优质服务。其正立面优美的轮廓，高低起伏富有韵味。整体造型一气呵成，表达了航空的"飞翔"意念，运用简洁的非线性金属屋面与曲面玻璃幕墙相结合，流畅精致。远期造型犹如飞行器一般，展开双翅，以一种大气、热情的姿态迎接往来旅客，其标志性、时代性和地域特征得到较好的平衡与协调。无论是建筑造型还是空间设计，潮汕机场航站楼都是一个根植于传统地域文化与气候特征，一个探索当代岭南建筑创新设计的原创建筑，一个集全新旅客体验、节能环保、高效运营需求于一体的独特的花园式建筑，其创新设计为我国干线机场航站楼的规划设计提供了有益启示。

Jieyang Chaoshan Airport is the third major airports in Guangdong following Guangzhou Baiyun International Airport and Shenzhen Bao'an International Airport. The new airport is of great significance for the region to realize traffic integration and economic integration and fulfill the objectives of "major changes in five years and major developments in a decade".

Chaoshan Airport Terminal is of a medium scale and is developed in three phases. The terminal in Phase I is a compound structure with concourses configured in a linear layout. Each of the three concourses leading to the main building is 200 meters in length, which not only is a proper walking distance for passengers but also helps create clear and convenient pedestrian flows. Both the main building and the concourses can be expanded in the future as needed. The interior design of the terminal building has drawn on the courtyard concept of traditional housings. A suspension garden is designed between the main building and the concourses. The garden facilitates the ventilation of the terminal and reduces energy consumption of the building with its architectural elements such as cornice, shading and local skylight.

Chaoshan Airport was completed and put into use in Dec. 2011. The terminal building's front facade incorporates the fascinating outlines of local housing in Chaoshan area, portraying an undulating yet charming building shape. The overall image of the terminal, being smooth and well coordinated, expresses the idea of "flying" in aviation. The concise non-linear metal roof is integrated with the curvilinear glazed curtain wall to create a flowing and exquisite appearance. It is a terminal building which is rooted in local culture and climate characteristics. The creative design of the terminal building has been an inspiration to designers who are involved in planning and design of terminal buildings of major airports in China.

项目地点：广东省揭阳市揭东县炮台镇
设计时间：2007—2010年
建设时间：2008—2011年
用地面积：380293m²
建筑面积：58752m²
建筑层数：地上3层，地下1层
建筑高度：30.17m
曾获奖项：国际竞赛第一名
2013年度广东省优秀工程勘察设计工程设计二等奖
2013年全国优秀工程勘察设计行业三等奖

Location: Paotai Town, Jiedong County, Jieyang, Guangdong Province
Design:2007-2010
Construction: 2008-2011
Site:380,293m²
GFA: 58,752 m²
Number of floors: 3 above-grade floors and 1 below-grade floor
Building height: 30.17m
Awards:
The First Place of international competition
The Second Prize for Excellent Engineering Exploration and Design of Guangdong Province (2013)
The Third Prize for National Excellent Engineering Exploration and Design Award (2013)

1　总平面图
Site plan

2　花园式航站楼概念草图
Conceptual sketch for gardenlike Terminal

3　航站楼与指廊之间的悬空花园，旅客在出发和到达的流程中可以欣赏到当中的景色，是潮汕机场空间特色之一
The overhead garden between the terminal and the pier, where passengers can enjoy the sceneries either during their departure or arrival; one of the spatial features of Chaoshan Airport

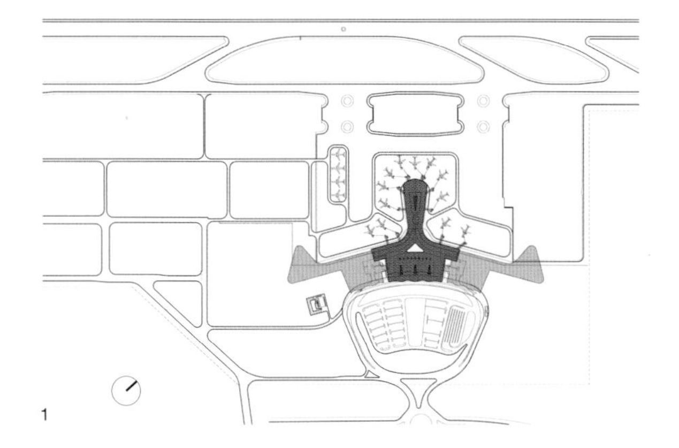
1

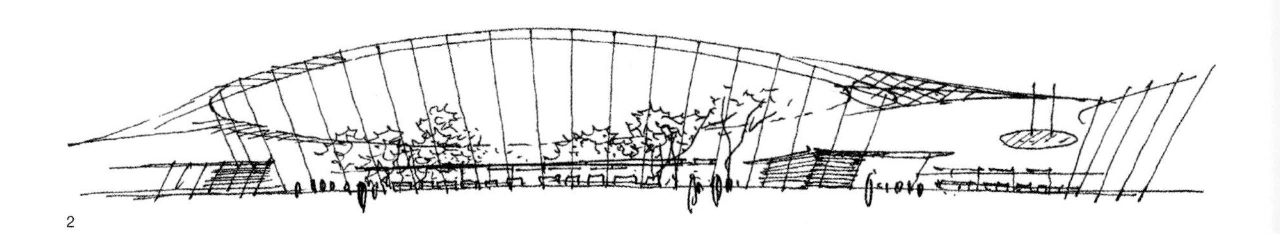
2

3

出发大厅入口，旅客可以看到航站楼优美起伏的屋顶
Entrance of departure hall, passengers can view the graceful wavy rooftop

28

1

2

3

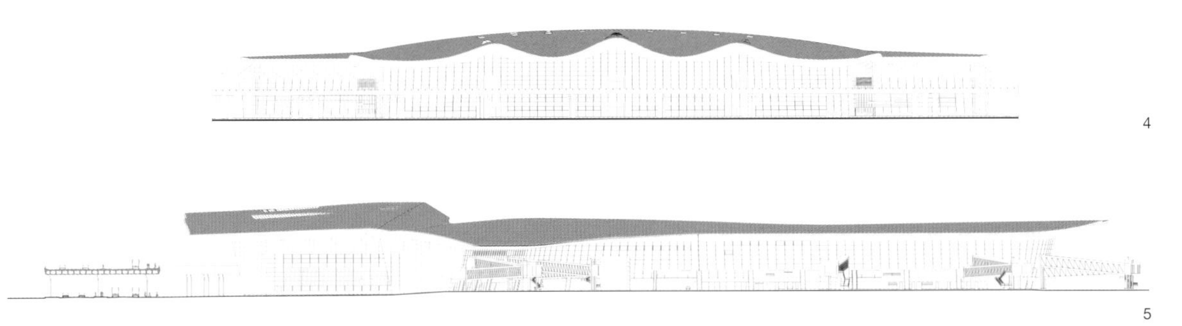

4

5

1 夜幕下的潮汕机场
The Chaoshan airport at night

2-3 中央指廊尽端适当放大，提供更舒适宽敞的候机空间
Appropriately enlarge the end of the central pier to provide a more comfortable and spacious waiting space

4 陆侧正立面图
Landside front elevation

5 侧立面图
Side elevation

1 透过候机大厅的玻璃幕墙看花园，幕墙外挂遮阳百叶，有效的调节室内光线
View into the garden through the glass curtain wall of the waiting hall; sunshade shutters are provided on the exterior of the curtain wall to effectively regulate the indoor light

3 行李提取大厅空间高大，视野开阔，尽享花园景色
Lofty baggage claim hall enjoys open and attractive garden view

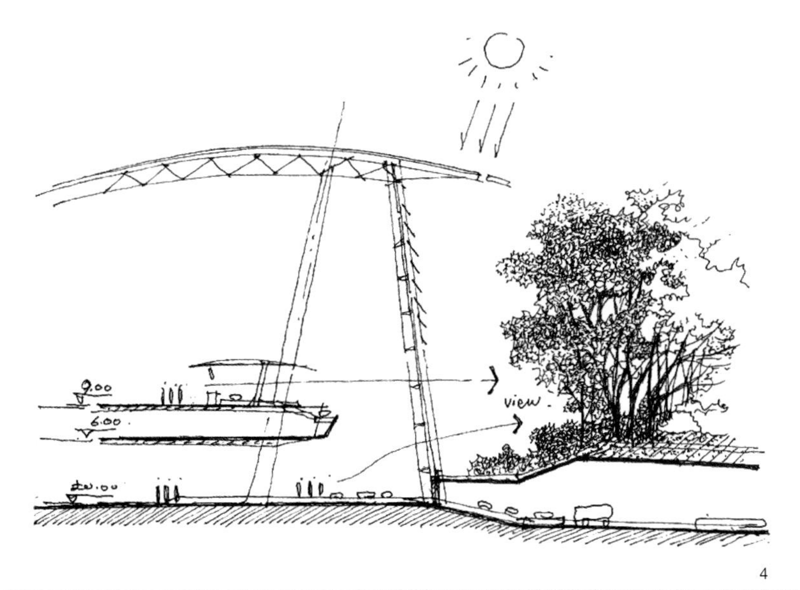

4

4 观景视线与日光处理手稿
Manuscript of viewing sight line and daylight treatment

2/5 花园促进航站楼的自然通风，结合飘檐、遮阳等手段降低建筑能耗
The garden promotes the natural ventilation of the terminal, and reduces building energy consumption in combination with the measures including roof overhung, sunshade, etc.

6 航站楼剖透视概念图：花园处于航站楼中心
Sectional perspective of the terminal concept: the garden is center of terminal

5

6

1

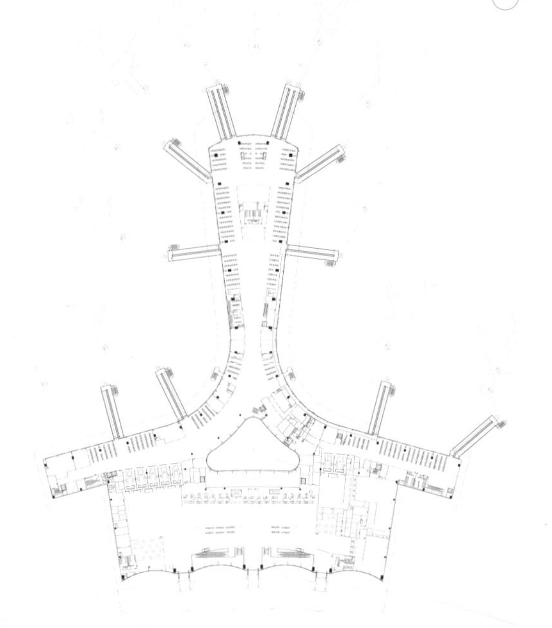

2

3

1 行李提取大厅
Baggage claim hall

2 三层平面图
Third floor plan

3 明亮通透的出发大厅，柜台背后是花园
Bright and transparent departure hall with garden at the back of counter

4 首层迎客厅
F1 check-in hall

5 利用自然采光的候机指廊
Daylit waiting pier

6 出发大厅的顶棚设计完美契合其上方的曲线屋面，天窗下方设有张拉膜顶棚，天光自然柔和
The ceiling design of the departure hall is a perfect match with the curve roof above. The tensile membrane ceilings beneath the skylights filter the soft daylight.

国际办票
International Check-in
29
-
39
值机柜台
Check-in Counters
纯净·新西兰 真爱 雅士利
出发

4

5

6

航站楼的造型表达了航空的“飞翔”理念，运用简洁的非线性金属屋面与曲面玻璃幕墙相结合，流畅精致。
The image of the terminal expresses the "flying" concept of aviation, and looks smooth and delicate by using concise non-linear metal roof in combination with the curve glass curtain wall.

武汉火车站
Wuhan Railway Station

2007-2009　湖北 武汉 / Wuhan Hubei

作为武汉市的门户建筑，武汉站犹如一只展翅的大鸟，寓意千年鹤归、九省通衢及中部崛起。其"大鹏展翅"的形象不仅映衬着"白云黄鹤"的历史文脉，更以展臂欢迎的姿态表达着城市的态度。武汉火车站于2009年12月26日建成启用。

武汉站位于湖北省武汉市杨春湖东侧，毗邻三环线，为武广客运专线湖北段的客运枢纽站及3个始发站之一（另外2个为新长沙及新广州）。客运站房规模庞大，建筑面积超过10万m^2，2020年每年旅客发送量将达到1750万人次，每日办理旅客列车162对，2030年旅客年发送量可达3100万人，高峰小时旅客发送量9300人。站内设有站台11座，共有20条股道，包括4条正线和16条到发线，是集高铁、公路、地铁、出租汽车、社会车辆于一身的交通建筑综合体，把空港的立体客运模式用于火车站设计，高效组织多种客运流线。武汉站首创等候式和通过式相结合的进站流线模式，"高架候车，上进下出"。旅客可选择进候车室候车进站，也可直接由绿色通道进站，更可在站厅俯瞰所有停靠于站台上的列车。武汉站实现了各种交通工具"零距离换乘"，形成铁路与城市公交、地铁的无缝连接，给乘客提供最大方便。该火车站还是规划中的城市轨道交通4号线、5号线的终点站。乘客不出站即可转乘地铁。以客运流线为主干，武汉站设计出丰富的内部空间，围绕巨大的中庭，两侧分布各区候车厅，室内良好的空间尺度感和光环境使得建筑本身舒适而节能。车站主体充分利用拱形结构的稳定性，并进行造型的艺术处理和构件的标准化设计，实现稳定的纵横双向连续拱跨体系,衍伸出"树状"钢结构和混凝土结构造型的核心元素。

武汉站不仅融入了丰富的文化内涵，更在技术和艺术上达到了统一，在使用功能和造型艺术上获得广大市民的一致认可，有幸成为武汉这座城市的"名片"。2012年，运营3年的武汉站获得芝加哥雅典娜建筑设计与博物馆颁发的"2012年国际建筑奖"，成为"世界最新最美的建筑"。

Wuhan Railway Station was completed and open for traffic on December 26, 2009. The architectural form presents a welcoming posture, showing a friendly attitude towards people from all over the world.

Located to the east of Yangchunhu Lake, Wuhan City, Hubei Province and adjacent to the Third Ring Road, the Station serves as the passenger junction station in Hubei section of Wuhan-Guangzhou High-speed Railway, as well as one of the three departure stations. Wuhan Railway Station is claimed to be one of the largest stations in China, having 11 platforms, a gross floor area of over 100,000m^2, and a total of 20 tracks including 4 main lines and 16 arrival and departure tracks. By 2020, the station will have an annual passenger throughput of 17.5 million people, with daily capacity of 162 pairs of passenger trains. The number is expected to rise to 31 million in 2030, with passenger throughput reaching 9,300 people at peak hour. The Station features a combined complex accommodating high-speed rail, highway, subway, taxis and other vehicles. Wuhan Railway Station has realized the zero distance interchange with various means of communication by providing seamless connection with bus and subway, thus bringing passengers the maxi-mum convenience. The Station also serves as the terminal of Lines 4 and 5 of the planned rail transport. With the platform provided at the first underground level, railway passengers enjoy the convenience of direct interchange with subway, eliminating the trouble of having to exit the railway station first. The Station is designed mainly based on passenger circulation service, providing vast interior space.

Wuhan Railway Station has not only incorporated rich cultural connotations, but also achieved the unity in architectural art and construction techniques. In 2012, three years after its operation, Wuhan Railway Station was awarded The International Architecture Award for 2012 by Museum of Architecture and Design, Chicago Athenaeum, and acclaimed as the newest and most beautiful building in the world.

项目地点：湖北省武汉市
设计时间：2003－2005年
建设时间：2007－2009年
用地面积：车站建筑工程建设总用地30.7hm^2
建筑面积：客运用房106841m^2
建筑层数：地上3层，地下2层
建筑高度：58m
合作单位：中铁第四勘察设计研究院集团有限公司（总包设计单位）
AREP公司（外方设计单位）
曾获奖项：2010年度铁路优质工程勘察设计一等奖
2011年评为中国"百年百项杰出土木工程"
2011年中国土木工程詹天佑奖
芝加哥雅典娜建筑设计博物馆颁发"2012年国际建筑奖"

Location: Wuhan, Hubei Province
Design:2003-2005
Construction:2007-2009
Site:30.7ha
GFA:10,6841m^2 for passenger transportation rooms
Number of floors: 3 above-grade floors and 2 below-grade floors
Building height: 58m
Partner:
China Railway Siyuan Survey and Design Group Co., Ltd. (General Contractor)
AREP (international firm)
Awards:
The First Prize for Excellent Railway Engineering Exploration and Design (2010)
"100 Outstanding Civil Engineering Projects from 1900 to 2010" (2011)
Tien-Yow Jeme Civil Engineering Prize (2011)
The International Architecture Award for 2012 by Museum of Architecture and Design, Chicago Athenaeum

透过吸音管帘天花可以看到一体化设计的主体结构、照明、吊顶系统。
Through the sound acoustic pipe curtain ceiling, the integrated design of the main structure, lighting and ceiling system can be viewed.

1

2

1 夜色下的武汉站，与湖面倒影相映生辉
A splendid night view of Wuhuan Railway Station and the lake

2 出发大厅立面"大鹏展翅"的形象
The façade of departure hall symbolizing "Roc on the ascent"

3 从剖面图可以看出建筑与结构的极度契合；站厅大空间、站台灰空间有序排列展开
The section shows the perfect integration between architecture and structure; the large space of station concourse and platform grey space spread rhythmically

4 剖透视展现"通过式"的候车方式
Sectional perspective, showing the through-type waiting style

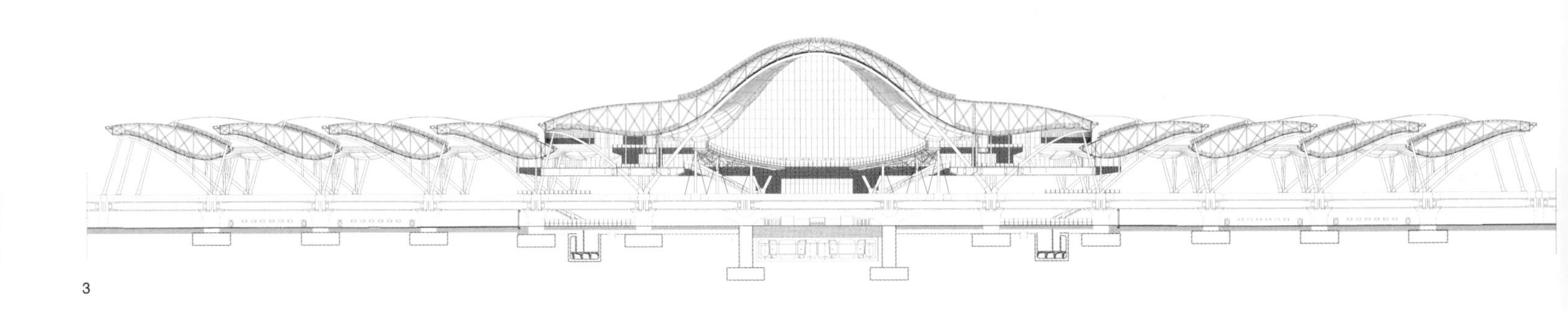

3

4

1 出发大厅：管帘吊顶有效调节车站内光线，节约运营成本
Departure hall: Pipe curtain ceiling effectively regulating the light ray in the station and saving operating costs

2 从出发厅可以俯瞰站台
Overlook the platform from the departure hall

3 武汉站出发大厅
Departure hall of Wuhan Railway Station

4 从上至下进站区、站台区、出站区、地铁区层次清晰，使用高效
Clearly-defined top-down hierarchy, i.e.entrance, platforms, exit and metro area contribute to high efficiency

1

2

3

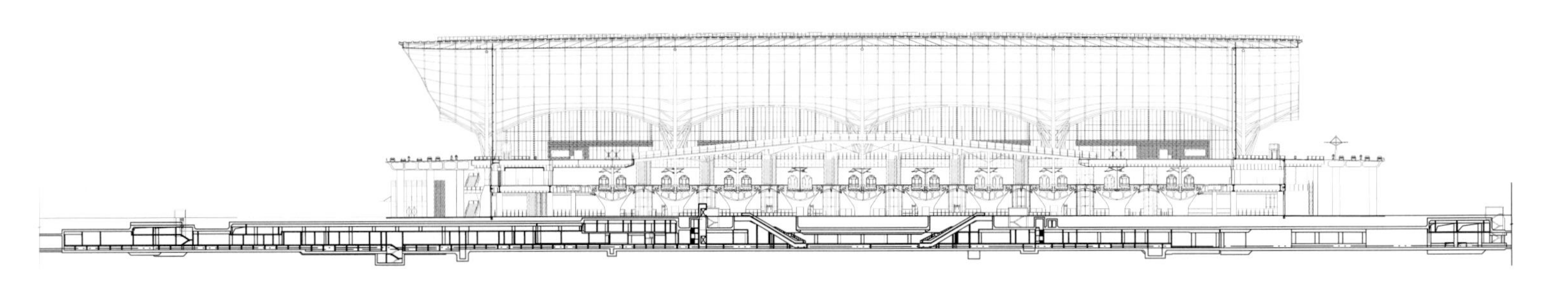
4

5

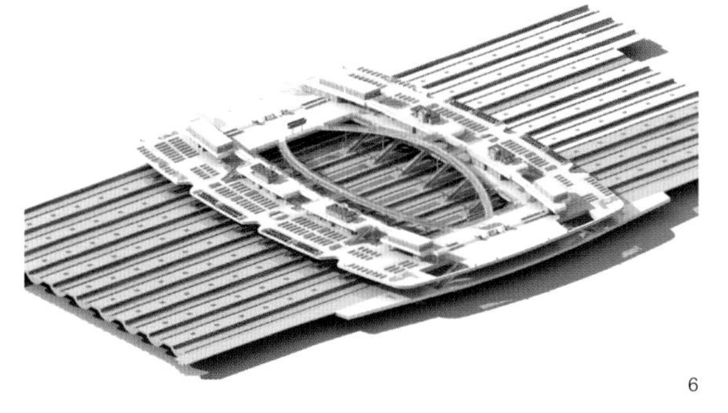

6

7

5 旅客候车空间：高侧窗及百叶吊顶引入了足够的自然光线
Passengers waiting space: Sufficient natural daylight introduced by high side window and shutter ceiling

6 武汉站内部功能布局模型
Internal functional layout model of Wuhan Station

7 吊顶系统及屋面体系与钢结构拱支撑融为一体
Ceiling system and roofing system are integrated with the supports of the steel structural arch

深圳机场新航站区地面交通中心（GTC）
Ground Transportation Center (GTC) of New Terminal Area, Shenzhen Airport

2009-2012　深圳 宝安 / Bao'an Shenzhen

随着航空业的快速发展，深圳宝安国际机场需要新建T3航站楼以满足机场吞吐量的增长。新T3航站楼承担起了空侧的交通运输，同时需要一座地面交通中心来解决陆侧复杂的交通运输。机场交通的特殊性、复杂性及多样性决定了作为主要交通枢纽的交通中心有着重要的地位。快速高效，以人为本，人员流程的清晰性和使用的便捷性是设计的宗旨。

项目为新T3航站楼的交通枢纽，为世界各地到港旅客提供各种交通和附加服务，是一个连接T3航站楼与航空城、轨道交通的一个多元化交通核心，承担着连接地上、地面和地下各类交通设施的任务。设计充分利用周边交通组织，将各类交通工具分流至交通中心各个部位，避免人流混杂。内部人流组织采用人车分流+平层换乘的方式，功能清晰，大大简化了复杂的交通情况，并且合理利用人流为建筑增加商业和餐饮附加值。交通中心同时也是一个"中心广场"，所有的商业和服务设施及各种交通模式都被集中在同一个屋檐的下面，相互交融构成了一个复杂的综合体。

GTC地面交通中心与T3航站楼是航空城的核心，两者设计浑然一体，GTC与T3航站楼的自由形态互相呼应，创造出独特的机场景观，以飞扬的形态向人们展示。GTC平面呈椭圆形，三维立体的双曲金属屋面、外倾斜的框式玻璃幕墙，建筑造型光滑、完整，形态沉稳、大气，充分体现"中心"地位。同时，设计充分利用自然元素，令建筑与环境和谐发展，通过天窗、玻璃幕墙，引入自然光线，充分利用周围自然景观，使建筑充满活力。

With the rapid development of aviation industry, it is necessary to have a new Terminal 3 in Shenzhen Bao'an International Airport to cope with the growing airport throughput. The new Terminal 3 is responsible to handle traffic and transportation in the airside; meanwhile, a ground transportation center (GTC) is also required to cope with the complicated traffic and transportation in the landside. The terminal design aims to realize fast, efficient and human-oriented operation, a clear passenger circulation and convenient use.

GTC is the traffic hub of the new Terminal 3, providing various transport and additional services to passengers arriving from all over the world. The design makes the best of spaces to organize traffic and distribute different transportation modes to different parts of the center, thus avoid mixing pedestrian circulations into other circulations. As for internal pedestrian organization, the pattern of "separated pedestrian/vehicle circulations + same level interchange" is employed. Moreover, it betters the commercial and F&B performances by attracting more people to this area.

GTC and Terminal 3 form the core of Aviation City, which are perfectly integrated as one in design. The building shape is smooth, complete, solemn and stately, fully embodying the significance of GTC as a "center". On the other hand, the design makes the best of natural elements to keep the building in harmony with its surrounding. The structural design of GTC pose a big challenge as the roof is connected with the outrigger of Terminal 3 while it is linked up with two metro lines underground.

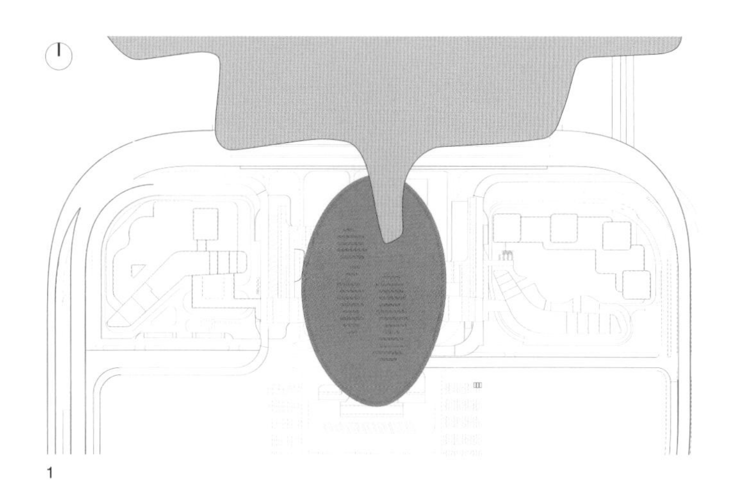

1

2

项目地点：深圳 宝安
设计时间：2009—2010年
建设时间：2010—2012年
用地面积：4.27万m²
建筑面积：5.8万m²
建筑层数：1-3层
建筑高度：27m
合作单位：意大利FUKSAS

Location: Bao'an, Shenzhen
Design:2009-2010
Construction:2010-2012
Site:42,700m²
GFA: 58,000 m²
Number of floors: 1-3
Building height: 27m
Partner: FUKSAS (Italy)

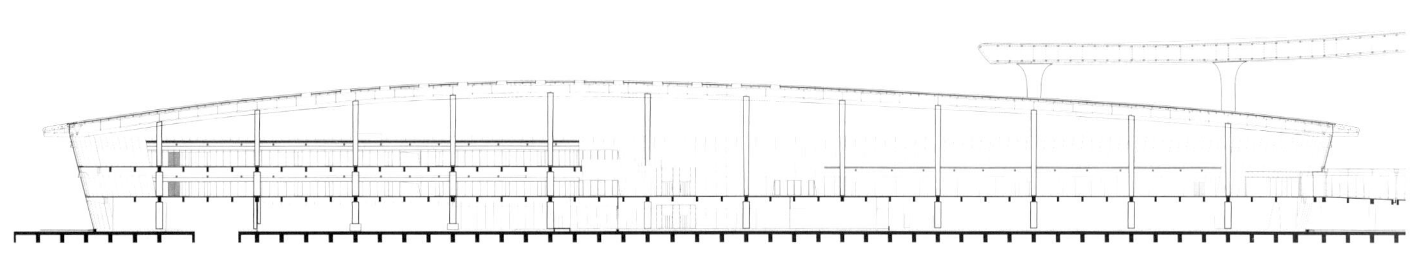

3

1 总平面图
Site plan

2 T3航站楼屋面伸臂与GTC屋面搭接
The terminal's roof reaches to splice with GTC roof

3 纵向剖面图
Longitudinal Section

1

2

3

1 从T3航站楼看GTC，自由形态相互呼应
A view of GTC from Terminal 3: the free forms echo to each other

2 GTC的外部装饰照明，主要通过室内的大空间照明辐射，透过通透的玻璃幕墙形成的，避免无功能的过度装饰。檐口底部设置连续光带，勾勒出建筑的轮廓，为城市增加一道美丽夜景
GTC's exterior decorative lighting is mainly realized by the lighting for interior large space penetrating through the transparent glass curtain wall, thus avoid the functionless excessive decoration. The continuous lighting strips along the bottom of the cornice portrays the building outline and contribute to the charming night view

3 GTC与T3航站楼是整体设计的两部分，无论在空中、楼层、地面均紧密相连
Being integral parts in the overall design, GTC and T3 are closely connected in air, on floor and at grade

1 与T3航站楼一致的"树形"风柱
Dendritic air columns same as those in T3

2 一层至二层的中庭，是人行流线上的关键交通节点；宽阔的中庭空间同时也是公共区的一个重要空间景观元素
The atrium from F1 to F2 is not only the key node along pedestrian circulation but also a key landscape element of public area

3 GTC二层换乘大厅，是GTC内部人流"平层换乘"的核心空间，白色的设计基调，使空间简洁、明亮、协调；六边蜂窝形顶棚形态独特而壮观；虚实变化的顶棚与条形采光窗紧密结合，创造出光影变幻的效果
GTC's transfer concourse on F2 is the core space of "exchange at the same level" for internal pedestrian, where the white design tone helps create a concise, bright and well-coordinated space featuring the unique and amazing hexagonal honeycomb ceiling; the illusionary change of ceiling and strip skylights jointly create the play of light and shadow

4 二层换乘大厅往南平台的中庭，聚集人流与商业结合
The atrium on F2's transfer concourse leading to the south platform gathers pedestrians into the retail area

1

2

3

4

二层至三层的中庭
Atrium from F2 to F3

P
3 号停车场
Park 3
公交
Bus

长沙黄花国际机场扩建工程
Design Competition for New Terminal Building Of Huanghua International Airport,Changsha

2007　湖南 长沙市 / Changsha Hunan

长沙黄花国际机场定位为国内干线机场和国际定期航班机场，是实现国家中部崛起战略和湖南"长株潭"区域经济发展的门户机场。设计方案以流畅的曲面造型象征飞翔理念，体现飞鸟般灵动的形态。航站楼屋顶上精心设计了大大小小的圆形天窗，夜间在内部灯光的照明下，波浪般起伏的屋面上面犹如繁星点点，象征美丽的"星城"长沙。

黄花国际机场新航站楼采用前列式指廊构型，北指廊布置国际机位、南指廊布置国内机位，指廊端部放大以停靠更多机位。在功能流线上采用出发和到达分层分流式。空侧的运行简洁高效，陆侧的交通及停车楼进深得到保证，总体布局做到空侧和陆侧的极佳平衡。航站楼的外形给人驭风而起的感觉——弧形的屋盖，舒展的指廊两翼，既似飞鸟翱翔，又如飞行中的机翼。飞鸟般灵动的形态，银光闪闪的外壳，具备了现代、流畅、浑然一体的建筑风格。主楼波浪般起伏的屋面，流畅多变的外墙，屋面和外墙并没有明显的分界线，变化莫测，步移景异，犹如一个不可分割的生命体。

航站区的设计方案从整体考虑，统一规划，分期实施。充分考虑与远期规划的衔接，并且预留一定的灵活性。实践证明，机场业务量的发展受到多种因素的影响。采用预留一定弹性和灵活性的方案，在以后发展中更有利于抓住机遇和抵御可能遇到的风险。新航站楼是机场总体规划的主体，其功能清晰，运营高效，服务灵活。本方案设计体现了当代建筑高科技的美感，为旅客提供一个独特而又便捷的出行和到达场所。新航站楼的设计都将为旅客和迎送客人的当地居民营造一个便利、舒适的环境场所，带来愉快而温馨的出行体验。

Changsha Huangsha International Airport is positioned to be a major domestic airport and an airport operating regular international flights. Moreover, it is a gateway airport that helps realize the strategy of Mid China rise and promotes the economic development of "Changsha-Zhuzhou-Xiangtan Region" in Hunan Province. The design scheme reflects the concept of soaring with free-flowing curved shapes, and embodies the agile image of a flying bird.

Round skylights in the roof symbolize the beautiful "Star City" Changsha. The new terminal of Huanghua International Airport features a "Pier + Linear Composition", where north pier is for international stands while south pier is for domestic stands. The two ends of the piers are enlarged to allow for more stands. The departing and arriving circulations are separated and arranged on different levels. Operation in the airside is clear and efficient. In the landside, transportation is carefully arranged and a sufficient depth is guaranteed for the garage structure. The two have achieved a perfect balance. As there are no obvious dividing lines between the undulating roof of Main Building and the free-flowing and changing external walls, the presence of the terminal building will change when viewed from different angles as if it were an integral living body.

The terminal area is subject to overall consideration, unified planning and phased implementation. The design considers both the connection with the long-term planning and the flexibility for future development. It has been proven that, since the airport development is subject to various factors, the design with certain elasticity and flexibility will better respond to future development in both opportunity and risk. As the core in the airport master plan, the new terminal building features clearly defined functions, efficient operations and flexible services. It creates a convenient and comfortable environment for local meeters and greeters at the airport, and brings pleasant travel experience to passengers including business people and tourists.

项目地点：湖南 长沙
设计时间：2007年
建筑面积：本期（2007年）规划年旅客量为1240万人次/年，建筑规模约9.5万m²；远期（2035年）规划年旅客量为2900万人次/年，建筑规模约25.8万m²。
建筑层数：3层
合作单位：美国兰德隆·布朗（Landrum Brown）公司
曾获奖项：2007年国际竞赛第二名
2007年广东省注册建筑师协会优秀建筑创作奖

Location:Changsha, Hunan Province
Design: 2007
GFA: This Phase (2007) - annual passenger throughput is 12.4 million person-times with a GFA of about 95,000m²
Long-term (2035) - annual passenger throughput is 29 million person-times with a GFA of about258,000m²
Number of floors: 3
Partner:Landrum Brown (USA)
Awards: The Second Place of international competition in 2007
Excellent Architecture Creation Award by Guangdong Chapter of Association of Chinese Registered Architects (2007)

2

1 总平面图
Site plan

2 黄花国际机场夜景效果图：航站楼弧形的屋盖给人驭风而起的感觉，如飞鸟翱翔
Night view rendering of Huanghua International Airport: The arc roof of the terminal looks like a flying bird in the wind

3 航站楼夜景鸟瞰图：精心设计了大大小小的圆形天窗，夜间在内部灯光的照明下，波浪般起伏的屋面上面犹如繁星点点，象征美丽的"星城"长沙
Bird's eye view of the terminal at night: Elaborately designed large and small round skylights, like stars dotted in the wavy roof at night when the indoor lights are on, representing the beautiful Star City - Changsha

4 航站楼低点透视：航站主楼波浪般起伏的屋面，流畅多变的外墙，屋面和外墙并没有明显的分界线，变化莫测，步移景异，犹如一个不可分割的生命体
Low view-angle perspective of the terminal: The wavy fluctuating roof and the smooth yet changeful façade share no distinct boundaries and look rather like an integral organism

5 正立面图
Front elevation

1

3

4

5

南宁吴圩国际机场新航站区及配套设施扩建工程

Design Competition for Terminal Building and Expansion of New Terminal Area and Supporting Facilities, Wuxu International Airport, Nanning

2009　南宁 吴圩镇 / Wuxu Town Nanning

随着广西壮族自治区和南宁市社会经济的快速发展，特别是建立中国东盟自由贸易区以来，机场航空业务量将迎来前所未有的高速发展期，机场航站区现有设施已不能满足其发展的需要，扩建既必要又迫切。南宁吴圩机场位于广西壮族自治区首府南宁市郊郁江南岸吴圩镇，距离南宁市区27.8km。目前场区周围已有较完善的地面交通网，利于周边城市旅客客源的快速集散。

新航站楼采用前列式+指廊式混合构型，其中南指廊及中间指廊提供22个近机位服务于国内航线，北指廊提供8个国际机位。三指廊的布局设计提供了较多的近机位及商业空间。指廊与主楼连接紧密，中央布置集中商业、旅客及CIP服务设施，在此处还布置了中央花园。航站楼跌宕起伏的曲面屋顶与远山的轮廓遥相呼应，再现了广西南宁山水如诗画般的意境；中央花园设计巧妙，贯通多层的露天花园位于楼中央出发和到达流线必经之处，出发与到达旅客均可强烈感受到南宁的"半城绿树半城楼"浓郁绿意；经过艺术提炼并简化富有地域特色的壮锦纹样，构成陆侧交通绿化带图案和屋面天窗肌理，阳光把壮锦纹样洒向航站楼室内地面，结合柔美的顶棚设计，让旅客体验到生动的光影变化。

飞鸟般灵动的形态，航站楼的外形给人驭风而起的感受。同时，建筑造型融入了航空器的特质银光闪闪的外壳，具备了现代、流畅、浑然一体的建筑风格。主楼波浪般起伏的屋面，流畅多变的外墙，犹如从环境优美的地平面乘势而起。飘动的曲线塑造出充满力量、速度和技巧的建筑动态，流动互融、节奏分明的建筑空间，给旅客深刻的印象，充分发挥机场作为城市门户的作用。

Being 27.8 km away from downtown Nanning, Wuxu International Airport is located in Wuxu Town in the suburban area of Nanning. Since the setting up of China-ASEAN Free Trade Area, the aviation business in Guangxi has witnessed an unprecedented high-speed development. Therefore, existing facilities in the airport terminal area were no long able to satisfy its development needs and expansion has become necessary and imperative.

The new terminal features a "Pier + Linear Composition", of which the south pier and the middle pier have totally 22 near stands that are for domestic flights while the north pier has 8 international stands. The layout of the three piers allows for more near stands and bigger commercial spaces. The ingenious design of the central garden has created a surrounding in which both departing and arriving passengers are deeply impressed by the lush green of Nanning which as the saying goes, is "half occupied by green trees and half occupied by buildings". The extremely characteristic patterns of Zhuang brocade have become the patterns in landside traffic greenery belts and the fabric of roof top skylights.

With a form that resembles an agile flying bird, the terminal looks as if it were ready to take off. Meanwhile, a distinctive feature of an aircraft is integrated into the building, creating a modern, fluid and integrated building. The floating curves outline a building that combines strength, velocity and workmanship; while the flowing, inter-connected and rhythmic building spaces highlight the airport as a main gateway to the city.

项目地点：南宁 吴圩镇
设计时间：2009年
建筑面积：12.9万m²
建筑层数：地上3层，地下1层
建筑高度：33.4m
合作单位：英国福斯特建筑事务所（Foster+Partners）
美国兰德隆·布朗（Landrum Brown）公司
曾获奖项：2009年年国际竞赛第二名
2011年广东省注册建筑师优秀建筑佳作奖

Location:WuxuTown, Nanning
Design:2009
GFA: 129,000m²
Number of floors: 3 above-grade floors and 1 below-grade floor
Building height: 33.4m
Partner: Foster+Partners (U.K.)
Landrum Brown (USA)
Awards:
The Second Place of international competition (2009)
Excellent Architecture Creation Award of Guangdong Chapter of Association of Chinese Registered Architects (2011)

1　本期（至2020年）航站区总体规划平面图
Master plan for terminal area (this phase till 2020)

2　剖透视图，中庭花园概念，出发和到达大厅旅客均可看到
Sectional perspective, atrium garden is visible from both departure and arrival hall

3　航站楼采用前列式+指廊式构型
The terminal features "linear + pier" composition

4　航站楼如飞鸟般灵动
Terminal looks like flying bird

5　出发办票大厅中心花园景观
Central garden of departure and check-in hall

6　迎客厅与出发大厅的空间连通
Spatial connection between arrival hall and departure hall

2

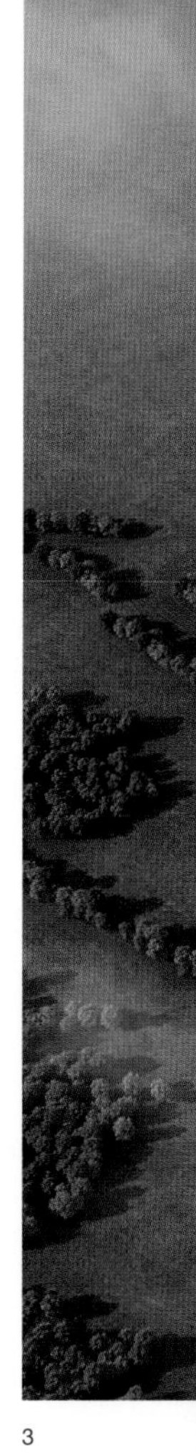

3

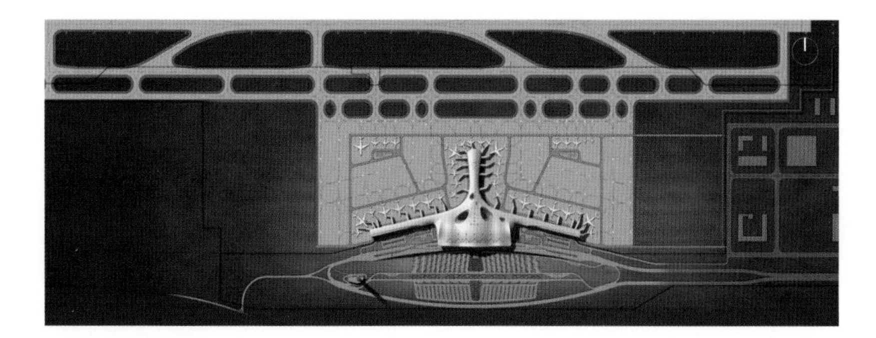

1

4

5

6

烟台潮水国际机场新航站楼方案竞赛
Design Competition for New Terminal Building of Yantai Tidewater International Airport

2011　烟台 潮水镇 / Chaoshui Town Yantai

2

本案要为烟台市创造一个达世界水平的航站楼，它承载传统建筑内涵，又是一座浪漫、高效、人文的现代化航站楼；一个现代设计手法和传统建筑精神实现最佳平衡的航站楼——用创新的设计手法演绎传统建筑意象；时尚设计结合回归自然精神的花园式航站楼——效率和休闲的有趣结合。

烟台潮水国际机场位于山东省烟台市潮水镇，靠近蓬莱市，距烟台市中心约43km。机场拟建设一条长3400m、宽45m两侧各设7.5m宽的道肩的跑道和等长的平行滑行道，同时新建航站楼8万m^2，拥有39个机位的停机坪以及总库容3000m^3的机场油库，总投资约40亿元。建成后，潮水机场旅客吞吐量将达到每年1000-1200万人次。

造型如流体般富有动感，延绵动态的曲线形成多层次的隐喻——山脉和海潮，显示出动感浪漫的设计意象。动感的航站楼飘檐以现代手法演绎传统建筑的飞檐，并为出发车道边营造出大量遮风挡雨的灰空间。屋面的波峰隐藏巨型结构，室内空间动感简约，支撑体系均成为空间或景观的有机部分，壮观夺目的花园设计贯通了出发和到达空间，成为画龙点睛之笔。

The design aims to create a world-class terminal in Yantai, which embodies the architectural connotations of traditional buildings and which is modern, romantic, efficient and human oriented.

Located in Chaoshui Town, Yantai, Shandong Province, Yantai Tidewater International Airport is close to Penglai and about 43km to downtown Yantai. It is planned to construct a runway which is 3,400m in length and 45m in width and has a shoulder of 7.5m in width on each side, as well as a parallel taxiway of the same length. Also, it is planned to have a new terminal of 80,000m^2, which consists of an apron of 39 stands and an airport oil terminal of 3,000m^3 in volume. The total invest is about RMB4 billion. Upon completion, the annual passenger throughput of the Airport will be 10 to 12 million person-times.

The new terminal features a "Pier + Linear Composition" layout, with high percentage of near stands. With fluid and dynamic building shape and the stretching and dynamic curves as metaphors, the upturned eaves of the dynamic terminal are a modern embodiment of the traditional overhanging eaves, considerable all-weather gray spaces for the departure curbsides.

4

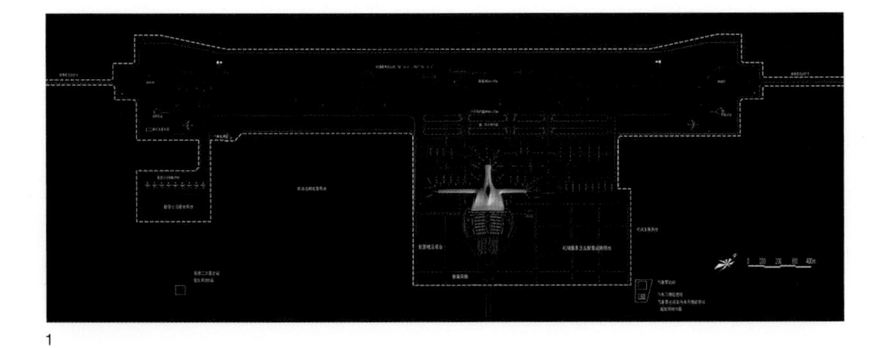

1

项目地点：烟台 潮水镇
设计时间：2011年
建筑面积：8万m^2
建筑层数：地上3层
建筑高度：33.8m

Location:Chaoshui Town, Yantai
Design:2011
GFA: 80,000m^2
Number of floors: 3 above-grade floors
Building height: 33.8m

3

1 本期（至2020年）航站区总体规划平面图
Master plan for terminal area (this phase till 2020)

2 鸟瞰效果图，造型如流体般富有动感，延绵动态的曲线隐喻山脉和海潮，显示出动感浪漫的设计意象
Bird's eye view: the fluid form and the flowing curves symbolizing mountains and tide present a romantic and dynamic image

3 中央花园与办票、安检空间融合
Central garden integrates with check-in and security check space

4 航站楼采用前列式+指廊式构型，造型流畅
The terminal in flowing form features a "linear + pier" composition

5 剖面透视图，出发流程围绕着中庭大花园有条不紊的展开，旅程感受美观且高效
Sectional perspective: smooth departure circulation centers on the atrium garden, offering nice view and efficient flow

5

广州亚运馆
Guangzhou Asian Games Gymnasium

2007-2010　广州 番禺 / Panyu Guangzhou

广州亚运馆（原名广州亚运城综合体育馆），是2010年广州亚运会唯一新建的主场馆，本届亚运会标志性建筑。亚运馆凭借其创新设计理念、独特的建筑体验、标志性和可实施性的绝佳平衡，在国际设计竞赛中中国原创获胜并且成为实施方案。

亚运馆包括了体操馆、综合馆和亚运博物馆等一系列功能。区别于传统体育场馆的设计方法，建筑师大胆地利用有机连续的金属屋面统领多个场馆，场馆犹如珍宝隐藏于屋面之下。场馆间、屋檐下的灰空间连贯舒展，体现了传统建筑文化和场所精神，大面积金属屋面轻盈飘逸，独具岭南建筑的神韵。建筑师注重营造积极开放的城市公共空间，设置穿越场馆的城市广场，为城市带来新活力。整个亚运馆造型生动，利用445R铁素体高耐候性不锈钢屋面板系统，与铝镁锰屋面板系统合成"双层皮"屋面系统，综合解决了造型美观及复杂屋面的排水问题，完美地实现设计团队期待的屋面曲线，从城市空间各个角度呈现出不断变化的造型以及色彩变化，展示亚运馆丰富的建筑形象。

亚运馆充分利用计算机三维模拟技术，完成钢结构、金属屋面板、玻璃幕墙及金属幕墙的设计。在材料、结构等方面进行了新的探索。例如：大面积使用隐藏拉索式双曲玻璃幕墙，通过整体分析为幕墙的构造、节点设计和连接设计提供精确的数据支持，使幕墙与主题结构完美结合；清水混凝土浇注面积达25000m^2，具有施工难度高、规模大、造型多样、高支模等特点，完成效果受到高度评价。同时，亚运馆倡导绿色亚运及可持续的设计理念，设计综合考虑到自然通风采光、太阳能照明及中水回收利用系统。

广州亚运馆设计与施工历时逾两年，跨越三载，对于一个如此复杂的大跨度三维连续曲面的体育建筑，这是巨大的挑战。在设计团队的共同努力下，仍然达到了极高的设计完成度，在新技术、新材料应用做了很多探索，在自主创新设计等方向实现了突破。随着项目竣工验收并投入测试赛，广州亚运馆以其梦幻般的形象和独特的体验迅速获得社会各界的高度关注，赢得赞誉！

Guangzhou Asian Games Gymnasium (the Gymnasium) is an iconic venue newly built for the Guangzhou Asian Games 2010. Our design proposal, with a wonderful balance between its symbolic appearance and practicability, won the 2007 international design competition and was finally implemented.

The Gymnasium consists of the Gymnastic Hall, the Indoor Stadium and the Asian Games Museum. A free-flowing metal roofing system interconnects the individual venues which appear to be treasures hidden underneath. The continuing grey spaces among the venues and under the eaves embody the traditional architectural culture and the sense of place. The mega metal roof, light and lofty, presents the unique charm and style of architecture in South China. By creating active and open urban public spaces and providing urban squares winding among the venues, the design injects vitality into the city and enhances the public attributes of the spaces. The roofing system comprising stainless steel panels of high weatherability tactfully tackles the complication with roof drainage, meanwhile contributes to the vivid, impressive and varied building images when viewed from different part of the city.

3D simulation technology is given full play in design of steel structure, the metal roofing system and the glazed/metal facade, while explorations are made on new materials and structures. For instance, hyperbolic glazed facade supported by hidden stayed cables are extensively used to perfectly integrate facade and the main structure. The 25,000m^2 of area finished with fair-faced concrete is well regarded upon completion despite of the unusual construction complexity/magnitude, varied shapes and high framework support. The natural ventilation, daylighting, solar lighting and reclaimed water collection and use are also incorporated into the design.

It takes just 2+ years to design and build the Gymnasium, which is a huge challenge for a sports venue of such complexity, large span and continuously waving cladding. Thanks to the concerted efforts of the design team, the Project is completed up to the desired effect with numerous explorations on new technologies/ materials and breakthroughs of independent innovation. Since its completion and operation, the Gymnasium is highly acclaimed for its dreamlike presence and unique spatial experience.

项目地点：广州 番禺
设计时间：2007—2008年
建设时间：2008—2010年
用地面积：101086.6m^2
建筑面积：65315.0m^2
建筑层数：4层
建筑高度：33.8m
曾获奖项：2008年国际竞赛第一名
詹天佑土木工程奖创新集体奖
2011年全国工程勘察设计行业优秀工程勘察设计行业一等奖
2011年荣获百年百项杰出土木工程
2011年度中国建筑金属结构协会颁发的中国钢结构金奖（钢结构工程结构设计秀奖）
2011年度广东省优秀工程设计一等奖
住房和城乡建设部2009年绿色建筑与低能耗建筑"双百"示范工程
2011年度第六届"中国建筑学会建筑创作优秀奖"
2011年度广东省注册建筑师优秀建筑创作奖
2011年度广东省空间结构学会颁发的广东钢结构金奖"粤钢奖"设计奖一等奖
2010年度China-Designer中国室内设计年度评选年度优秀公共空间设计金堂
2011年评为中国第十届"詹天佑土木工程大奖"
2012年第六届中国建筑学会建筑创作优秀奖
2013年香港建筑师学会两岸四地建筑设计大奖优异奖
AAA2014亚洲建筑师协会奖：专业建筑类别荣誉奖

Location: Panyu, Guangzhou
Design:2007-2008
Construction:2008-2010
Site: 101,086.6 m^2
GFA: 65,315.0m^2
Number of floors: 4
Building height: 33.8m
Awards:
The First Place of international competition(2008)
Tien-yow Jeme Civil Engineering Prize (Innovation Team Prize)
The First Prize for National Excellent Engineering Exploration and Design - Engineeri Exploration and Design (2011)
"100 Outstanding Civil Engineering Projects from 1900 to 2010" (2011)
Gold Prize for Steel Structure (Excellence in Steel Structure Design) by China Construction Metal Structure Association (2011)
The First Prize for Excellent Engineering Design Award of Guangdong Province (201
Model Project for Two 100-Top Green Building and Energy Saving by Ministry of Housing and Urban-Rural Development of PRC (2011)
Excellent Award for the 6th ASC Architectural Creation Award (2011)
Excellent Architecture Creation Award of Guangdong Chapter of Association of Chinese Registered Architects (ACRAGD) (2011)
The First Prize for Gold Prize for Steel Structure Guangdong Provincial Society for Spatial Structures by Guangdong Provincial Society for Spatial Structures (2010)
Jin Tang Prize for China Interior Design Awards 2010 – Excellent Public Space Desig
The 10th Tien-Yow Jeme Civil Engineering Prize (2011)
Excellent Award of the 6th ASC Architectural Creation Award (2012)
Merit Award, Cross-Strait Architectural Design Awards 2013 by Hong Kong Institute Architects
The ARCASIA Award (AAA2014): Honor Award – Architecture

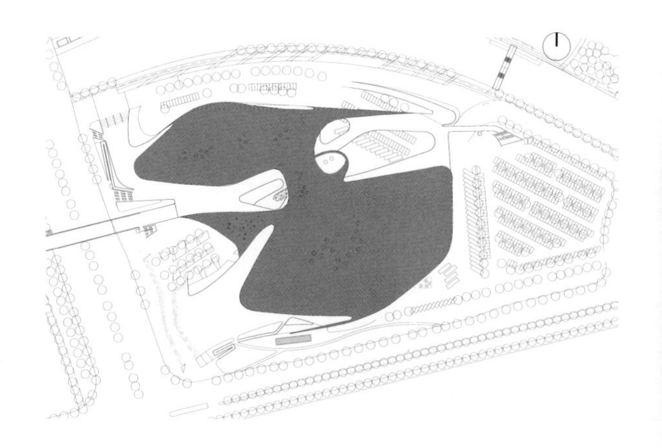

总平面图
site plan

有机连续的金属屋面外壳统领各个场馆。
The organically continuous metallic shell unifies various venues.

1

2

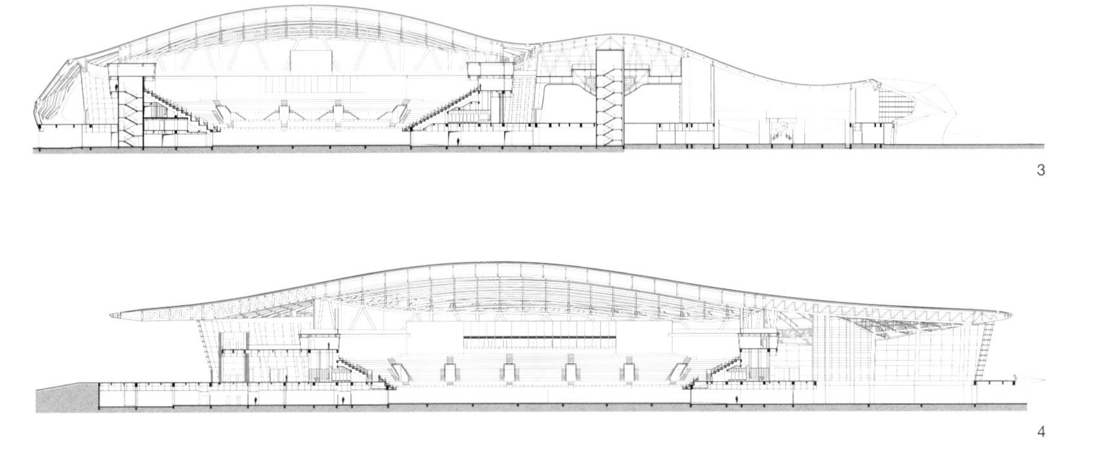

3

4

1 有机连续的金属屋面外壳统领各个场馆
The organically continuous metallic roof shell unifies various venues

2 亚运馆概念设计手稿
Conceptual design sketch of Asian Games Gymnasium

3-4 亚运馆剖面图
Section of Asian Games Gymnasium

1

2

3

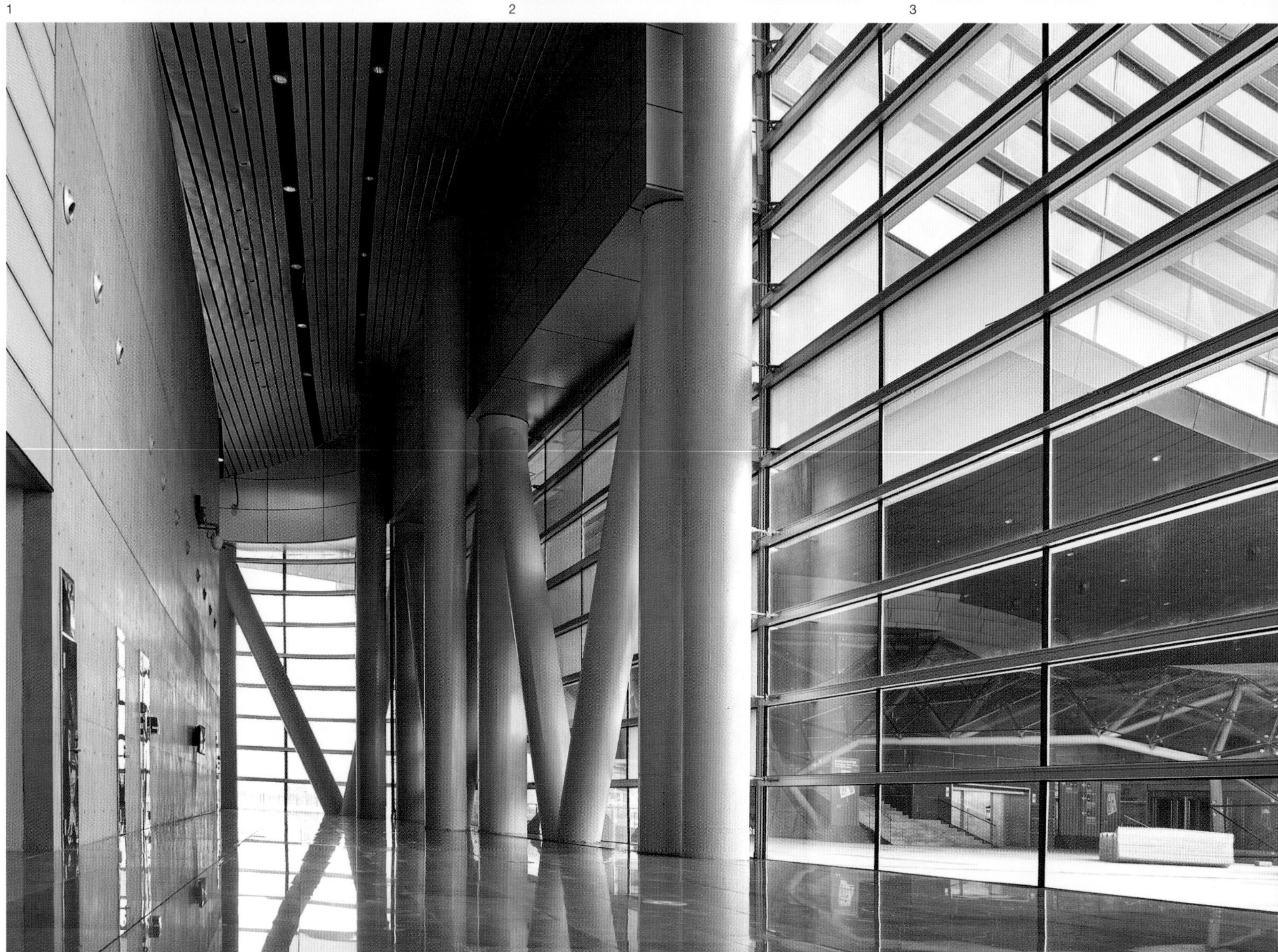

4

1 设计方案大胆地利用有机连续的金属屋面统领多个场馆，场馆犹如珍宝隐藏于屋面之下
The design boldly employs organic continuous metal roof to integrate multiple venues, which are just like jewelries concealed under the rooftop

2-3 曲线屋面下的灰空间给参观者类似传统建筑檐下的空间感受
The gray space under the curve roof presents visitors a spatial experience of standing under traditional building eaves

4 流动的室内空间与清水混凝土墙
The flowing interior space and bare concrete wall

5 亚运历史展馆入口采用结晶体驳接爪件点式幕墙，造型独特
The entrance to the Asian Games History Pavilion employs crystalline connected with the claw point-type curtain wall, featuring a unique image

6 从二层平台灰空间看亚运历史展馆入口
View of entrance to the Asian Games History Pavilion from the grey space on the terrace of F2

7 亚运馆空间节点设计手稿
Spatial node design sketch of the Asian Games Gymnasium

8 受当地传统建筑的影响，亚运馆飘檐出挑大，呼应了该区域的气候，从而降低建筑能耗
Asian Games Gymnasiums have large roof overhung, echoing with the climate in this area and thus reducing the building energy consumption

5

6

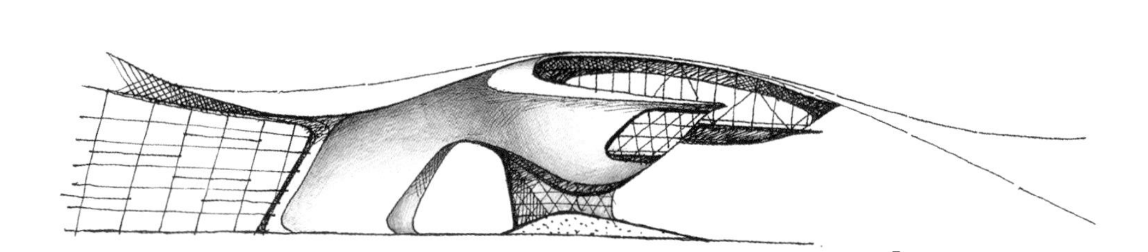

7

7

1 亚运馆入口细部设计手稿
Detail design sketch for the entrance of the Asian Games Gymnasium

2 入口广场上方设有天桥与位于二层的公共平台连接，鼓励人群从地铁站步行进入场地，为城市空间带来新活力
An overpass is provided above the entrance plaza to connect the public platform, encourage people to walk to the site from the metro station and inject new vitalities into urban space

3 金属屋面系统在解决屋顶排水的同时，提供了完美的建筑造型
The metal roofing system solves the roofing drainage problem while providing perfect architectural image

4 三维屋面营造了流动连贯的灰空间
3D rooftop creats a flowing continuous grey space

5 比赛场馆室内
Interior of competition venues

2

4

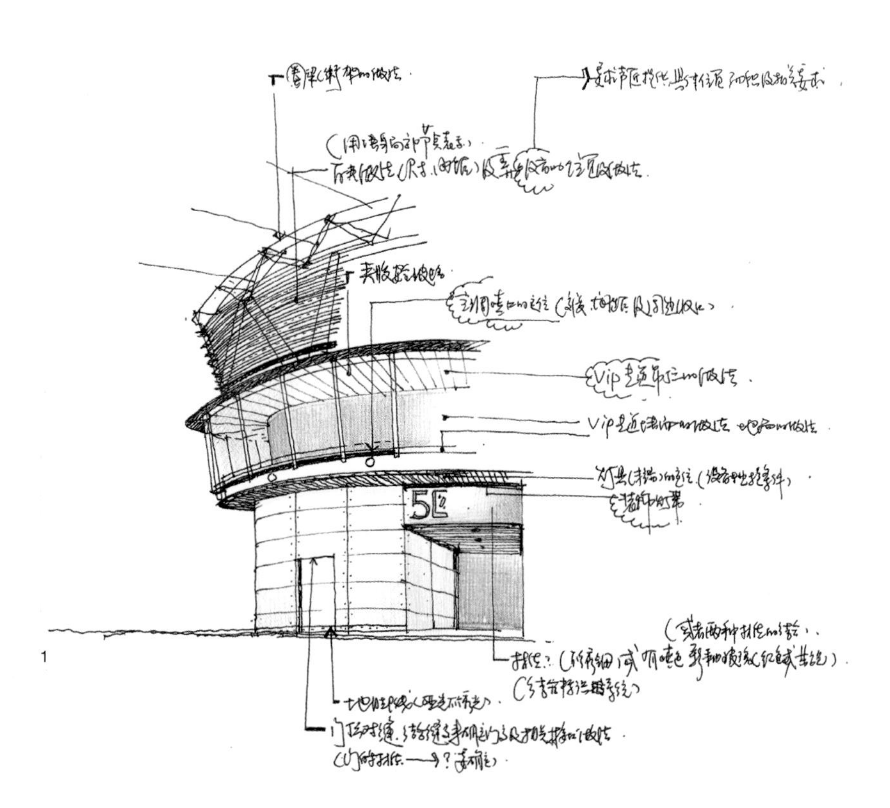

1

3

5

夕阳下的亚运馆
Asian Games Gymnasium at sunset

广州亚运主媒体中心
Main Media Center, Asian Games Town, Guangzhou

2007-2010　广州 番禺 / Panyu Guangzhou

亚运城主媒体中心是2010年广州亚运会唯一的大型媒体建筑，主要功能为新闻中心和国际广播中心，是一个能够满足各国记者对各项赛事进行采访、转播、报道、联络等需求的综合设施。亚运城主媒体中心与广州亚运馆（原名：广州亚运城综合体育馆）作为一个整体，在2007年的国际设计竞赛中以中国原创获胜并且成为实施方案。主媒体中心位于亚运城西南部，西临轨道交通4号线，东跨官涌并通过人行空中漫步廊与广州亚运馆相连，北侧为媒体人员居住的大型媒体村，南边为预留发展用地。

亚运城主媒体中心方正的建筑体量配上流动柔美的造型线条，变化丰富的侧墙和边角，增加了空间的戏剧性；精致的玻璃幕墙开窗结合文字的标示性，以及鲜明颜色的铝板造型，彰显出媒体建筑的个性，为媒体工作者创造了富有激情的工作平台。内部空间围绕中庭展开，同样富有动感。南侧人行空中漫步廊造型丰富，曲线优美，有移步换景的效果。廊桥长约500m，最大跨度为53m，桥面宽度为12—33m，有机地把广州亚运馆、主媒体中心及轨道交通4号线海傍站连成一个整体。

亚运城主媒体中心是2010年广州亚运会为媒体人员服务的综合性建筑，赛后将改建为大型商业中心，服务整个广州亚运城片区。

Guangzhou Asian Games Main Media Center is the only large-scale media building for the 2010 Guangzhou Asian Games and a comprehensive facility that meets both domestic and foreign journalists' demands for interviews about, transmission of, reporting on and liaison about various games. The design of GAGT Main Media Center and Guangzhou Asian Games Gymnasium as a whole won out as an independent Chinese entry in the International Design Competition in 2007 and finally became an implementation plan. It is located in the southwest of Guangzhou Asian Games Town (GAGT), with the Haibang Station of the Metro Line No. 4 to the west, the media village to the north, the reserved land for development to the south, and Guanchong Creek to the east. It is connected with Guangzhou Asian Games Gymnasium via an overhead pedestrian corridor.

The square volume of the Main Media Center, coupled with gently flowing shape lines and varied side walls, edges and corners, adds to the dramaticism of the space. The interior space is expanded from around the atrium with rich dynamism. The sky corridor for pedestrians in the south features varied forms and graceful curved lines and offers changing views with each step. The corridor, which is 500m long and 12 ~ 33m wide with the largest span at 53m, links the Guangzhou Asian Games Gymnasium, the Main Media Center and the Haibang Metro Station into an organic whole.

The Main Media Center will be redeveloped into a large commercial center to serve as a comprehensive commercial building for the entire GAGT area after the Games.

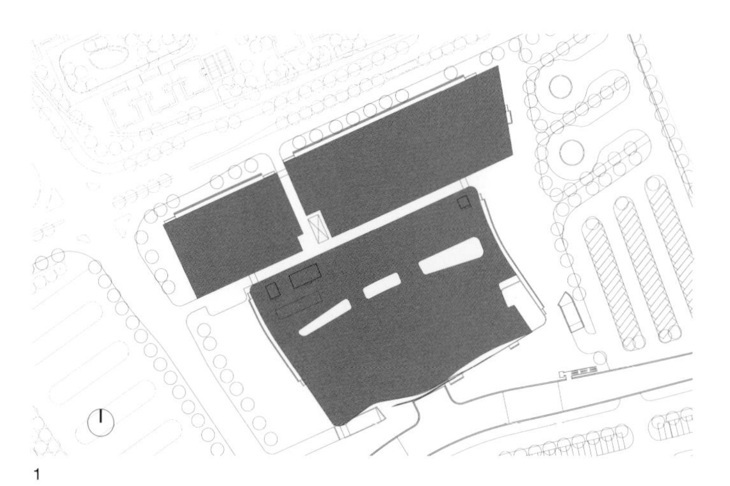

1

项目地点：广州市番禺区亚运城
设计时间：2007—2009年
建设时间：2008—2010年
用地面积：101969m^2
建筑面积：60012m^2
建筑层数：4层
建筑高度：23m
曾获奖项：2007年国际竞赛第一名
2011年广东省优秀工程设计二等奖

Location: Guangzhou Asian Games Town, Panyu District, Guangzhou
Design: 2007-2009
Construction: 2008-2010
Site:101,969m^2
GFA: 60,012m^2
Number of floors: 4
Building height: 23m
Awards:
The First Place of international competition(2007)
The Second Prize for Excellent Engineering Design Award of Guangdong Province (2011)

1　总平面图
Site plan

2　立面上有醒目的广州2010亚运会的标识，凸显了建筑的特殊性
The conspicuous signage of Guangzhou Asian Games 2010 is provided on the façade, highlighting the distinctiveness of the building

3　主媒中心与亚运馆效果图：主媒中心从二层伸出漫步廊桥，与亚运馆、轨道交通连接成整体，构筑更便捷的交通系统
A walking bridge extending out from F2 connects the Asian Games Gymnasium and the rail transit into a whole, creating a convenient traffic system

2

3

1

2

1　主媒体中心体量方正，但整体造型运用了流动柔美的线条，变化丰富的侧墙和边角，增加了空间的戏剧性
The Main Media Center uses square massing with flowing and mellow lines; changeful sidewalls and rim corners further enhance the dramatic feel of the space

2　室内中庭
The Atrium

3　首层平面图
Ground floor plan

3

广州花都区东风体育馆
Dongfeng Gymnasium, Huadu District, Guangzhou

2008-2010　广州 花都 / Huadu Guangzhou

"青山翠微迎露珠，秋意新雨霁绿藓。"——花都区东风体育馆采用简洁圆润的椭圆形，与其西南侧的康体公园和谐共生，与青山相互映衬，以独特的建筑形象与空间特点成为当地地标。本项目是2010年广州亚运会期间新建的体育场馆之一，也是花都中心城区西部的文化体育中心。体育馆能容纳8000名观众。满足体育场馆赛时及赛后的多功能使用是建筑总体设计的核心价值。

在权衡业主的需求、资金、施工周期等综合因素后，我们放弃了非线性的自由建筑形态，以简单实用为基调，采用最纯粹的几何形体量处理方法，以完整的体量与开阔的空间和周边建筑取得平衡。在椭圆体型基础上，金属屋面板、玻璃幕墙以及玻璃雨棚共同描绘出飞扬动感的曲线，整体具有自由流畅的动态形象。不规则的侧窗、天窗在打破简单外观形象的同时，为室内空间带来灵动、通透的光影。从第五立面看，体育馆又成了含苞待放的花蕾。

体育馆可以概括为两大功能空间——比赛馆和训练馆。主从有别的两个体量由曲线平台连接，形成自由、流畅的统一整体。比赛馆内观众休息厅与比赛大厅相互联通，在有限的结构空间中实现建筑使用空间的最大化，创造出扩大化的视觉感观及使用效率，从而节省面积。

在满足大型国际体育赛事的复杂使用要求的同时，充分考虑全民健身需求及不同公众活动的需求。亚运后，场馆成为该区域的全民健身运动中心，也作为举办各种大中型活动（演唱会、展会）以及汽车城工业产品的展销平台等得以有效利用。

结构设计借用了"箍桶原理"，在国内首次创造性地设计出了环形管内预应力大跨度钢结构体系。其结构构件小，安全性高，是同类体育场馆用钢量最低的，最大限度地节约了投资造价。

The olive-shaped Gymnasium harmonizes with the amenity park in the southwest and echoes to the mountains. As one of the newly built sports venues for 2010 Guangzhou Asian Games, the Project also functions as the cultural and sports center in the west of Huadu downtown. With the capacity of 8,000 spectators, the design accommodates the multi-purpose demand during and after the game, thus create a landmark with distinctive image and spatial characteristics.

We adopt the concise geometric volume to harmonize with the surrounding buildings with complete volume and open space featuring plain and practical style. Based on the olive shape, the metallic roof, glass curtain wall and glazing canopy jointly contribute to the light, free, smooth and dynamic image. Viewed from the 5th façade, the Stadium looks like a budding flower.

The Gymnasium is a large volume composed by two functional spaces, i.e. competition pavilion and training pavilion, which are free and smoothly connected by curved platform. The spectator lounge hall is connected with the competition hall in the competition pavilion, which maximizes the usable area in the limited structural space, expands the space visually and enhances the efficient, thus save the area.

In addition to the demand of large international sports events, the design has also fully considered the demand of public fitness and public events. After the Asian Games, the Gymnasium serves as the regional public fitness center.

Referencing the "wooden barrel principle", the structural designer creatively proposes the pre-stressed large span steel structure system in ring tube for the first time. Thanks to the smaller structural components and high safety performance, the steel consumption is the lowest among the similar stadiums, which significantly cut the investment cost.

项目地点：广州市花都区
设计时间：2008年
建设时间：2008—2010年
用地面积：73000m²
建筑面积：31416m²
建筑层数：4层
建筑高度：33.3m
曾获奖项：2008年国内竞赛第一名
广东省第六次优秀建筑佳作奖
2011年度广东省优秀工程设计二等奖
2011年度广东钢结构金奖"粤钢奖"
2011第十四届中国室内设计大奖赛学会奖
2011年度全国工程勘察设计行业优秀工程勘察设计行业三等奖

Location:Huadu District, Guangzhou
Design:2008
Construction: 2008-2010
Site:73,000m²
GFA: 31,416 m²
Number of floors: 4
Building height: 33.3m
Awards:
The First Place of national competition(2008)
The 6th Excellent Architecture Creation Award by Guangdong Chapter of Association of Chinese Registered Architects
The Second Prize for Excellent Engineering Design Award of Guangdong Province (2011)
Gold Prize for Steel Structure of Guangdong Province (Yue Gang Award) (2011)
Society Prize for the 14th China Interior Design Awards (2011)
The Third Prize for National Excellent Engineering Exploration and Design (2011)

2

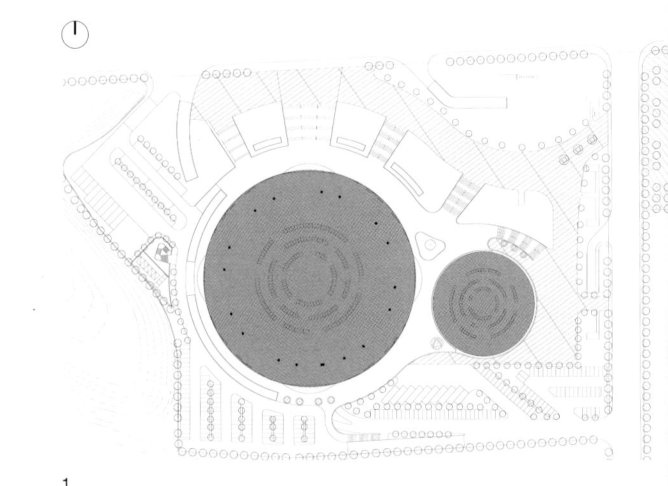

1

3

4

5

1 总平面图
Site plan

2 体育馆采用简洁圆润的椭圆形，如两颗露珠落在落在自然景观之中
The Gymnasium employs concise and oval like building forms resembling two dewdrops within the natural setting

3 体育馆概念手稿
sketch of Dongfeng Gymnasium

4 体育馆南立面
South Façade of the Gymnasium

5 体育馆建造过程模拟
Construction process simulation of the Gymnasium

1

2

1 体育馆夜景
Night Scene of the Palaestra

2 体育馆局部设有尺度适宜的下沉庭院
The Gymnasium is partially provided with appropriately-sized sunken courtyard

3 通向后勤服务区的通道及二层平台的大台阶
The large steps leading to the back-of-house areas and the F2 terrace

3

1

2

3

1 体育馆门厅的灯光设计简洁明了，凸显其空间特色
The lighting design in the vestibule of the Gymnasium is concise and clear-cut, highlighting the feature of the space

2 屋面钢结构斜立柱
Inclined column of the roof steel structure

3 体育馆门厅
Vestibule of the Gymnasium

4-5 场馆内部
Interior of the venue

4

5

惠州市金山湖游泳跳水馆
Jinshan Lake Swimming and Diving Complex, Huizhou

2006-2010　广东 惠州/ Huizhou Guangdong

惠州市金山湖游泳跳水馆是广东省第十三届省运会场馆之一，设有满足国际标准赛事及残运赛事的游泳池、跳水池、热身训练池、训练室及室外训练池。游泳跳水馆在非赛事阶段对公众开放，并作为当地运动员日常训练场地。以其鲜明新颖的建筑形象和极具感染力的室内空间，为市民提供了一处现代体育与文明科技高度结合的运动场所，也为城市营造了一处具有标志性意义的公共活动空间。以建筑与自然的合奏为形态构思的出发原点，巧妙利用丘陵、水与山谷地状特殊性，建筑体布局呈平缓起伏状。建筑体在青山衬托下呈现出多姿多彩的表情。设计以"山水意象"为主题，阐述环境和建筑的关联意义。

建筑物的外形既像一座座起伏的山峦，又如运动员在水中搏击的阵阵波浪。这一设计理念不仅仅体现在其流动的屋顶造型上，更加配合了该体育建筑的功能与空间的需求。游泳跳水馆内跳水池、比赛池、训练池为一字形布置，在功能上比赛与训练的区域分开，在空间上则连在一起，使馆内的空间更加宏大。而流动的屋顶造型所营造的空间，增加了场馆内部的层次感和趣味性。三维造型的跳台及椭圆形的玻璃背景板设计独一无二，也是馆内空间的焦点。运动员从里面走出来，就像登上一个竞技的舞台，一个人生的舞台，一个开启梦想的舞台。

我们采用了新现代主义风格的自由建筑形态，在形体交接处理上采用非线性手法，在建筑内部空间上，则强调钢结构的韵律美。通过组合变形手法，通过材质对比处理，线条丰富而流畅，形体简洁纯净又不乏细部。以明晰的构成逻辑表现了复合的建筑内容，以清新的手法表达了对体育文化的理解，使建筑显现出一种清雅、灵动的气质，呼应游泳跳水馆的使用功能。场馆屋盖结构纵向和横向均呈流线型，由横向布置的平面刚架、撑杆，纵向布置的次梁，斜向布置的钢拉杆和屋面系统组成。由于钢结构屋盖纵向高度变化较大，为了减小风及温度的纵向作用对结构造成的不利影响，结构设计借鉴了桥梁设计的成功经验，通过合理地选用固定支座、单向活动支座以及多向活动支座等新型抗震减振支座，合理地释放了温度作用，减小了结构构件的尺寸，既经济又满足了建筑师对建筑整体效果的要求，使得室内结构看起来更为轻巧、飘逸。

As one of the venues for the 13th Guangdong Provincial Games, Jinshan Lake Swimming and Diving Complex is equipped with the swimming pools, diving pools, warming-up pools, training rooms and outdoor training pools that meet the standards of normal international games and the para games. During the non-competition period, the Complex is open to the general public. The Complex offers a sports venue perfectly combining the modern sports with the technology and a highly representative public activity space.The design creates gently undulating building layout. The Complex presents diversified appearance against the green mountains. Themed on "picturesque mountains and water", the design elaborates the relation between the environment and the Complex.

The Complex appears like the undulating mountains, or the waves generated by the competing swimmers in the pool. The diving pool, competition pool and training pool are provided in a line. The competition and training area is separated functionally yet connected spatially to allow for generous space. The unique multi-layer diving platform and olive-shaped glazing background are the main attraction, which offer a stage for the athletes to go for the dream.

We adopt Neo-modern and free architectural form and non-linear design approach for building connection, while highlighting the attractive rhythm of steel structure in the interior space. With combination and variety and comparison between materials, the design presents diversified and smooth lines and concise yet detailed form. The vertically and horizontally streamlined roof consists of transversal planar steel frames and supports, vertical secondary beams, diagonal steel tie rods and roofing system. The structural design references the bridge design approaches to downsize the structural components, making the interior structure appear lighter and more graceful.

项目地点：广东省惠州市
设计时间：2006—2007年
建设时间：2006—2010年
用地面积：32440m²
建筑面积：24574m²
建筑层数：3层
建筑高度：28.8m
曾获奖项：2006年国内竞赛第一名
2011年第六届中国建筑学会建筑创作佳作奖
2011年全国优秀工程勘察设计行业奖二等奖
2011年广东省注册建筑师优秀建筑佳作奖
2011年广东省优秀工程设计二等奖
2011年惠州市优秀工程设计一等奖

Location:Huizhou, Guangdong Province
Design: 2006-2007
Construction: 2006-2010
Site: 32,440m²
GFA: 24,574m²
Number of floors: 3
Building height: 28.8m
Awards:
The First Place of national competition(2006)
The 6th ASC Architectural Creation Award (2011)
The Second Prize for National Excellent Engineering Exploration and Design (2011)
Excellent Architecture Creation Award by Guangdong Chapter of Association of Chinese Registered Architects (2011)
The Second Prize for Excellent Engineering Design Award of Guangdong Province (2011)
The First Prize for Excellent Engineering Designs of Huizhou (2011)

1

1　设计充分发挥金属屋面系统的优势，墙面、屋面一气呵成
The design gives fully play to the advantage of metal roofing system, which gets the wall and rooftop done without any letup

2-3　金山湖游泳跳水馆造型柔和起伏，与周边环境相协调
Jinshan Lake Swimming and Diving Complex features a soft and wavy building form in concert with the ambient environment

2

3

1

2

3

4

1-2 游泳馆的内部结构强调了钢结构的韵律美和简洁美，利用规则的结构体系创造出自由变化的三维曲面空间
The interior structure of the swimming hall highlights the rhythmic and concise esthetics of the steel structure, using regular structural system to create a free-changing three-dimensional curve space

3 跳台背景墙设计独特，给运动员提供一个展示实力的"舞台"
The background wall at the diving platform is concise and generous, offering athletes a stage to show their strength

4 在屋面高低错动之处设有侧窗，为室内引入自然光线
In addition to the wavy rooftop, side windows are provided to introduce natural daylight into the interior

游泳跳水馆正立面
Front Façade Of The Swimming And Diving Complex

样游泳比赛
竞赛门票处
泳館

宜昌奥林匹克体育中心概念规划及建筑设计
Conceptual Planning and Architectural Design for Yichang Olympic Sports Center

2013　湖北 宜昌 / Yichang Hubei

宜昌是巴楚文化的发源地，拥有2400多年悠久历史，以山水文化闻名。作为长江三峡的门户，宜昌是乘船游江的中途枢纽港，是众所周知举世闻名的名胜观光地。设计团队充分解读宜昌的“历史·文化·自然”，提出以“让山水、人、都市活力跃动”为主题，打造将环境、都市、教育、娱乐、商业融为一体的“体育生态公园”，希望以此为契机，培育优秀运动员和振兴宜昌市体育文化，推动宜昌向拥有更高层次的“经济·环境”都市发展。

设计方案综合考虑地块内河流、丘陵等地貌元素，建筑与周边环境和谐共生：四座大型体育场馆——体育场、体育馆、羽网运动中心、游泳馆，沿河岸展开，形成“发展轴”；射击馆则融合到地块北面的丘陵地带中，以减少噪声对周边环境的影响。一个位于二层的“滨水漫步廊”沿“发展轴”方向把体育场馆都连接起来，与带状生态区交错，营造出具有独特场所精神的城市广场。“滨水漫步廊”被设计成如长江般滂沱的流水形态，造型圆润的体育场馆如碧波之上的涟漪，点缀于漫步廊之上。体育场馆屋顶起伏，形成具有流动形态的天际线。用地北面平坦地段则布置了户外体育场地和设施。沿河岸一带修筑亲水公园，提供市民享受自然，野餐露营的场所。

奥林匹克体育中心内还布置有商业设施、酒店等功能，形成具有多样化功能的体育公园。各种场馆、设施与滨水漫步廊连接，各功能区联系便捷。赛时与赛后，体育竞技与全民健身、市民休闲等使用需求在这里都得到满足。整个体育中心绿意盎然，充满朝气，成为让体育健儿展示实力的舞台，同时也是市民在此闲庭信步，欢度时光的城市开放空间。

Based on the full understanding of Yichang's "History, Culture and Nature", the design team has proposed to use "Bring vitality to Landscape, people and urban space" as the theme to create a "sports ECO park" that integrates environment, city, education, entertainment and business.

The design proposal gives due to consideration to the river, hills and other topographical elements, trying to make the buildings stand in harmony with the surrounding environment: the four major sports venues - stadium, gymnasium, badminton and tennis sports center and swimming pool will be arranged along the river bank, forming the "development axis", while the shooting hall will be situated in the hills to the north of the site in order to reduce the impact of noise on the surrounding environment. A waterfront promenade located on the second floor will connect the sports venues in the direction of the "development axis". The area to the north of the site will provide outdoor sports space and facilities. A waterfront park will be built along the river bank to offer a place for the public to enjoy nature, camping and picnic.

The Olympic Sports Center will also be equipped with commercial facilities, hotels and other functions to create a sports park with diversified functions. The whole sports center constitutes a green, lush and vibrant space, offering a stage for athletes to showcase their skills and strength. It also provides a great urban open space where people may walk, rest or spend some time for pleasure.

项目地点：湖北省宜昌市
设计时间：2013年
用地面积：95万m^2
建筑面积：25.4万m^2
建筑层数：6层
建筑高度：50m
合作单位：日本佐藤综合计画

Location:Yichang, Hubei Province
Design:2013
Site:950,000m^2
GFA:254,000m^2
Number of floors:6 floors
Building height:50m
Partner:AXS

1

1　各个场馆犹如流淌在碧波之上展现的“涟漪”，呈柔软的圆形点缀在“滨水漫步廊”上
Like the ripples on the water, the venues are gently dotted along the "waterfront promenade"

2　滨水漫步廊把几大场馆联系起来，与带状生态区交错。河道沿岸修筑亲水公园，提供市民亲近自然的场所
The waterfront promenade connects several venues and interweaves with the banded ecological zone. Waterfront park is built along the river course to provide citizens with the paces to get close to nature

2

1

2

1 体育场外沿上设置了中央大厅，将成为市民的交流空间。沿途看到的景观不断变化，可体验移步异景的乐趣
The central hall at periphery of the sports venues will offer the social space for citizens. The views keep changing as people moves around

2 在2层高度布置一个大平台，将之命名为"滨水漫步廊"，利用这个平台沿发展轴方向将各个设施连为一体
A large platform is provided on F2, i.e. "waterfront promenade" to link up various venues along the development axis

3 体育场效果图
Rendering of the stadium

3

广州西塔
Guangzhou West Tower

2004　广州 / Guangzhou

项目地点：广州 天河
设计时间：2004年
用地面积：31084m²
建筑面积：492926m²
建筑层数：91层
建筑高度：388m
合作单位：日本原广司＋Atelier.phi建筑设计研究所
曾获奖项：2004年国际竞赛优胜方案之一

Location:Tianhe, Guangzhou
Design:2004
Site:31,084 m²
GFA: 492,926m²
Number of floors: 91
Building height: 388m
Partner: Hiroshi Hara ＋Atelier.phi
Awards:Shortlisted proposal

正值当今中国日新月异的大发展时期，广州市新城市轴线的形成将在全世界的瞩目下以实施推进。广州珠江新城西塔的建设将是新世纪高层建筑的里程碑！一座建筑史上前所未有的高层建筑，一座实现人类"天空之城"梦想的建筑，一座生态、节能、可持续发展的建筑，一座充满活力、聚集人气、高商业回报的建筑，是我们对西塔设计的提案。

从古到今，从东方到西方，一直流传着"天空之城"的梦想，从古代的嫦娥奔月、天上人间、仙女下凡、巴比伦通天塔，到现代的太空探索，人类一直都梦寐以求能够离开地面，建立空中城市。今天，这个梦想将通过这座城市中心的超高层建筑，在最新科学技术的驱使下变成现实。

地面以上的西塔由三大部分构成，分为南、北两栋的办公楼，之上是穹顶人空间式的超五星酒店，穹顶顶部则是观光厅。建筑空间的构成完全打破了传统高层建筑的模式，包含了一系列构成丰富的公共空间，地下空间的地面化，裙房的玻璃龙商业街，观光厅、酒店共享空间、扶梯之路都将吸引大量旅游、观光、购物、饮食的客人。高档次、灵活多变的办公空间，为不同需求的租户提供了最大程度的选择余地。夜幕降临，华灯初上，西塔灯火璀璨，随着办公楼灯光渐渐熄灭，上空浮现出一个梦幻般的"天空之城"。创造出世界上独一无二的酒店景观，超五星酒店的地位宛若天成。人们从首层乘透明电梯直上观光厅，穿过"天空之城"，恰似飞上云天。广州市美丽的都市风景尽收眼底。从观光厅乘扶梯迂回而下，仿若仙女下凡，犹如置身于童话世界。从不同高度眺望城市景色，让人流连忘返。

月球表面风化岩孔洞的地表肌理给我们许多设计灵感。总平面的设计，在地面和地下楼板开设的孔洞光庭，由高大榕树围成的广场庭园，其构思均源于此。构成了许多大小不一的城市广场，也为广州市民创造了新的"城市客厅"。竞赛委员会组织专家，对所有提交方案进行了评审，按照技术文件的要求，评出了五个优胜方案，日本原广司＋Atelier.phi建筑设计研究所＋广东省建筑设计研究院联合体的方案名列其中。

In the prime of China's full-speed development, the formation of the new city axis of Guangzhou will be pushed ahead under close attention of the world. The development of the West Tower in Zhujiang New Town, Guangzhou will mark a milestone in the history of high-rise buildings in the new century.

Since ancient times, there have always been dreams for a "sky city". Today, this dream will come true under the drive of the latest scientific technologies through this skyscraper at the city center.

The above-ground structure of the West Tower is composed of three parts: the south and north office towers and the capacious dome resting thereon which accommodates a super five-star hotel. The composition of the architectural spaces overturns the pattern of traditional high-rise buildings, incorporates a series of diversified public spaces, brings the underground spaces to the ground, and designs a shopping street in the shape of a glass dragon in the apron. The high-class, flexible office spaces offer the broadest spectrum of choices for tenants with different demands. When the night falls and the lights turn on, the West Tower will glitters against the sky. With the lights in the office towers gradually going out, a dream-like city appears in the sky. This is the one-of-the-kind view in the world. People take the transparent elevator at the ground floor to the observation hall, like ascending to the sky, to have a panoramic view of Guangzhou. While taking the escalators down from the observation hall like fairies descending on earth, people will find themselves in a fairyland, where they are captivated by the panoramic view of the city from different heights.

Our design is inspired by the weathered craters on the moon. The daylight courtyards form urban squares of varied sizes thus offer new activity spaces to the local citizens. For the design competition, five proposals are shortlisted by the expert panel under the Competition Committee according to the design brief, including the one submitted by us in collaboration with Hiroshi Hara Atelier.phi as a consortium.

1　区域日景鸟瞰图
Regional bird's eye view in daylight

2　从观光厅乘扶梯迂回而下，仿若仙女下凡，犹如置身于童话世界。从不同高度眺望城市景色，让人流连忘返
Visitors descending from the panoramic hall via the escalators will find themselves indulging in a fairytale world, where the urban sceneries viewing from various heights can all be enjoyed

1

2

1

2

3

1 室外广场与下沉广场相连，使地下空间地面化
The outdoor squares are connected to the sunken squares, transforming the underground spaces into above-grade ones

2 下沉广场、入口柱廊与"玻璃龙"商业体构成丰富的建筑景观
The sunken squares, entrance colonnade and the "glass dragon" retail facility jointly create a diverse architecture landscape

3 以"差异性"创造出新的"双子塔"。独特造型的西塔，被平行与新城市轴线的平面切割，以剖面作为东立面，犹如镜子，对东塔产生了"对面性"的孪生关系
New twin-tower is born out of diversification, The west tower with unique image is cut by the plane that is in parallel with the new urban axis; the section then becomes the east facade, generating a twin-relationship like a mirror with the opposite east tower.

4-5 酒店空中大堂
Hotel's sky lobby

4

5

东莞市商业中心区F区（海德广场）
Dongguan Commercial Center Zone F (Hyde Plaza)

2005-2013　广东 东莞 / Dongguan Guangdong

本项目的设计理念是创造"城市之门"与"没有背立面的建筑"的建筑形象，用"包容"的形象与城市景观和谐共融，刚柔并济，塑造东莞独一无二的五星级商务酒店和高档的写字楼，使其成为东莞的标志性建筑。建设地点位于东莞市新城市中心区，是东莞市商业中心区（又名：东莞第一国际商业区）的收官项目。其东北面为会展中心，西南面为购物中心，周边规划有轻轨站连接广州和深圳，地段得天独厚，

项目是一个集超五星级城市商务酒店、甲级写字楼、餐饮、会议、商业等功能为一体的大型建筑综合体，是一个顶部连体的超高层双塔建筑。以地块中轴线展开布局，建筑的中心位置正好位于原建筑群中轴线，总体对称，浑然一体，并以对称的姿态与地块正对面的会展中心相互呼应，建筑与城市空间具有广泛的共融性。"双Y形"塔楼布局避免相互之间的视线干扰；建筑各种功能集约高效、合理分流。建筑立面以"对称性"和"趣味性"相结合的姿态出现，以"对称性"对应于建筑群的轴线与整体城市空间，使建筑群与城市空间有很好的衔接并具有更广泛的关联性；在对称的建筑体量中做了很多不对称的立面设计，以"趣味性"对应于商业建筑功能的多样性和复杂性。

场地以中央圆形广场为"太阳"中心，各景观元素及设施以"光环"形式向外顺序扩展。酒店大堂四层通高，面向中心花园广场，曲线形的空间轮廓，通透的玻璃幕墙，开阔的大堂共享空间、充分彰显出五星级酒店的超然气度。Y形平面拥有最小的核心，最大的采光面，标准层空间舒适紧凑，使用率高。塔楼在不同高度设有2～4层的空中庭园，由此产生复式花园套房，以适合高端客人的需要。办公区设置的空中庭园可以作为休憩空间，提高办公环境质量，实现酒店房型和办公空间的多样化。本工程结构属A级高度的抗震超限工程，连体结构采用弧形钢结构桁架，共两榀，分别与内塔楼南、北两侧的外框柱连接，实现"城市之门"的建筑形象。

The Project intends to create an inclusive architectural image that features "gate of the city" and "building without back facade". The building, in its complete form, will rise as a unique five-star business hotel and high-end office building and also a landmark in Dongguan. The Project is favorably located in the new urban center of Dongguan City as the final anchor of Dongguan Business Center, with the Convention and Conference Center to the north-east, shopping mall to the southwest and light rail station planned in the vicinity to connect Guangzhou and Shenzhen.

The Project is designed to create a large building complex that combines super five-star business hotel, grade A office building, restaurant, conference, commerce and other functions. Moreover, the introduction of dual Y tower layout will avoid mutual visual interference. The various functions of the building are arranged in an intensive and efficient manner and reasonably distributed.

With the central circular square as the "sun" center, various landscape elements and facilities will be configured sequentially in the form of "halo". The façade of the building is symmetrical and interesting at the same time. It is symmetrical because of its relation with the axis of the building cluster and the whole urban space, which makes the building cluster and urban space well connected and more extensively associated. It is interesting because of the diversity and complexity of the commercial building.

项目地点：广东 东莞
设计时间：2005－2008年
建设时间：2006－2013年
用地面积：5.12万m^2
建筑面积：21.7万m^2
建筑层数：地上37层，地下2层
建筑高度：158m
曾获奖项：2005年国内竞赛第一名

Location: Dongguan
Design:2005-2008
Construction: 2006-2013
Site:51,200m^2
GFA: 217,000m^2
Number of floors: 37 above-grade floors and 2 below-grade floors
Building height: 158m
Awards:The First Place of national competition(2005)

1　方案设计手稿图
Sketch of schematic design

2　总平面图
Site plan

3　"东莞之门"的城市意象
Image of "Gateway to Dongguan"

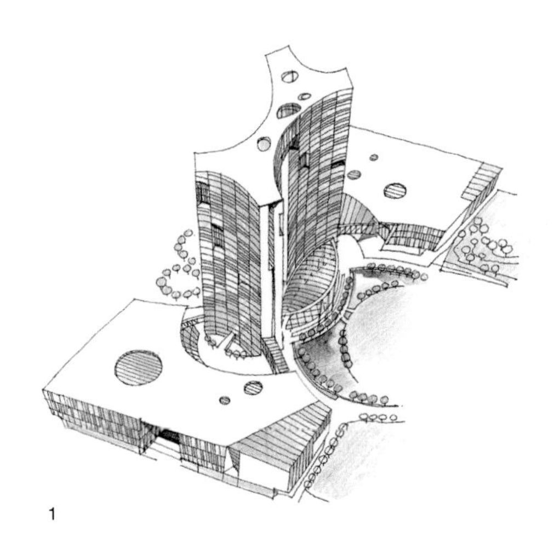

1

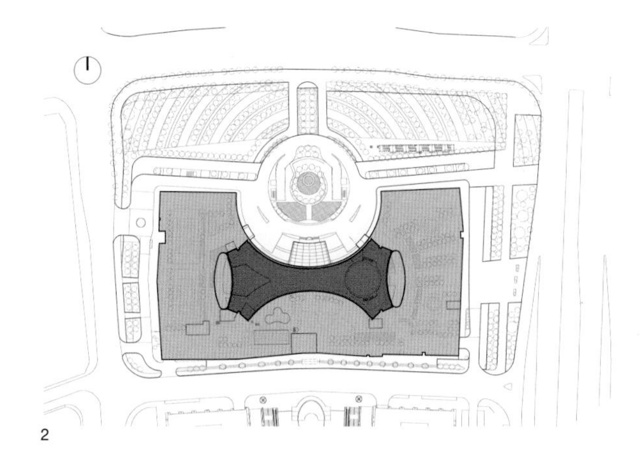

2

1

2

3

1 酒店大堂楼梯
Hotel lobby stairs

2 酒店侧庭
Hotel side foyer

3 酒店大堂
Hotel lobby

4 酒店主入口
Hotel main entrance

5 设计手稿图
Design sketch

4

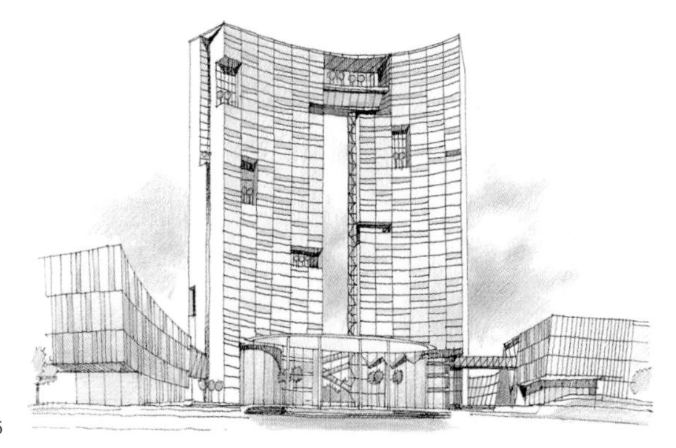

5

6

7

8

6 低点日景图
A day view form low viewing point

7-8 玻璃幕墙表皮
Glass curtain wall skin

夜幕中的海德广场
Hyde Plaza at night

香港新福港地产·佛山新福港广场
SFK·SFK Plaza, Foshan

2010— 佛山 禅城 / Chancheng Foshan

3

无论从建筑规模，还是建筑功能方面，新福港广场无疑是佛山的城市地标，以及区域中心。它是一个集客运站、商业、住宅功能于一体的大型城市综合体，总建筑面积约为54万m²，建筑高度接近180m。

新福港广场位于城市主干道魁奇一路、汾江南路的交汇处，也是佛山地铁1号、2号线路的交接处。其地下二层至四层为设备用房及停车库，地下一层是地下商业和连接地铁的交通空间。地上裙房则包含了客运站和城市商业空间。裙房中央设有贯通首层至四层的弧形中庭，引入自然光，并把各个功能区域串联起来，是商业空间的核心所在。裙房上方设有九座塔楼。由于用地条件紧张，根据项目及市场策划要求，合理设计塔楼高度及朝向，使每座住宅都能最大限度的满足当地严格的日照及消防间距等规划条件，因而也营造出丰富变换的城市天际线。

新福港广场是一个基于城市公共交通规划（TOD-transit oriented development）的城市综合体项目，它集合了地铁、公交车、出租车等公共交通系统，为商业开发带来充足的客流，促进商业发展。同时，强调高容积率的住宅开发，适当调整住宅户型的面积比例，提供更多的户型和户数，使核心区土地价值得以最大化利用。

The Project undoubtedly serves as the urban landmark and regional center in terms of size and functionality. This 180m tall urban complex is planned with a GFA of about 54 0,000m² and integrates the functions of coach terminal, commercial center, residence and apartment.

The project is prominently located at the intersection of Kui Qi Yi Lu and Fen Jiang Nan Lu and the Foshan Metro Line 1 and 2. B1 is used for retails and the traffic space connecting metro station. The above-grade podium is planned with coach terminal, large public transportation hub and commercial spaces. A four-floor-height curve atrium is provided in the center of the podium as the core of commercial space, introducing the daylight and linking up all functional zones. The podium roof equipped with the club and swimming pool serves as the floating garden for the residents of night residential buildings and apartment tower above the podium. The transportation systems of tower and podium are completely separated form each other.

As an urban complex based on TOD (transit oriented development), the Project attracts sufficient consumers to facilitate commercial development. The land value of the core area has been maximized by increasing FAR of residence, adjusting the proportion of residential area and providing more residential units.

项目地点：佛山市禅城区
设计时间：2010年至今
建设时间：2013年至今
用地面积：81101m²
建筑面积：550770m²
建筑层数：51层
建筑高度：180m
合作单位：Aedas环球凯达设计事务所

Location: Chancheng District, Foshan
Design: 2010 to date
Construction: 2013 to date
Site:81,101m²
GFA: 550,770m²
Number of floors: 51
Building height: 180m
Partner: Aedas

1 四层露天庭院效果图。露天庭院引入商业空间内，活跃了商业气氛，同时又使顾客得到不同的体验
Rendering of the outdoor courtyard on F4. The courtyard is introduced into the commercial space to foster active commercial atmosphere while offering different experiences to the customers

2 三层中庭庭院模型图。错落有致的中庭空间，利用屋顶光棚引入自然光，令商业空间丰富起来，更能提升商业气息
Diagram of the central courtyard on F3. The well-arranged courtyard introduces natural light through the rooftop skylight canopy to enrich the commercial space and enhance the commercial atmosphere

3 挺拔修长的体形更具冲击力
The lofty shape brings stronger visual impact

4 宽阔的休闲集散广场、通透明亮的玻璃盒子，增添不少商业气息
The spacious leisure square and transparent glass boxes enhance the commercial atmosphere.

5 塔楼丰富的天际线、错落有致的幕墙；裙楼石材与玻璃的虚实变换、部分退台的手法，加上入口造型的点缀，营造出浓郁的现代商业气息，体现出城市新地标
The diversified skyline and staggered façades of the tower, the void-solid alternation of stone and glass as well as the setback approach at partial positions of the podium, coupled with the unique shape of the entrance, foster strong atmosphere of modern commerce and create a new landmark in the city

5

1

2

4

罗浮山悦榕庄酒店
Banyan Tree Resort, Mt. Luofu

2011—　广东 惠州 / Huizhou Guangdong

悦榕庄酒店坐落在广东罗浮山脚，依山面湖，景色秀美，是全球顶尖精品度假村、公寓住宅及SPA的营运商——悦榕庄酒店集团的系列产品之一。因此，设计上秉承了该集团经营理念——为宾客提供寻求心灵回归与内在平静的处所。

悦榕庄酒店设有约100间别墅客房，每间客房独门独户，有山景别墅、园景别墅、水岸别墅等类型。整体建筑设计沿袭岭南的地方传统及地域特色，建筑布局以传统广府村落（渔村）为原型，利用起伏错落的地势，同时考虑最大限度的保留利用场地原有的植被，建筑依山傍水，几栋别墅为一组，围绕水体形成一个个小村落，简朴之意内外相同，安谧静好。建筑采用了传统民居形式，从小青瓦屋顶、青砖墙、起翘的屋脊，到客房内的精致装饰、家私、摆设，都体现了岭南传统风味与特色。

酒店设施完备，设有会议接待区、餐饮区、水疗区、客房区、运动区等功能区域，提供丰富多样的休憩活动空间。另外，酒店内还设有企业村专为团体客人提供便捷舒适的住宿服务，同时拥有悦榕Spa水疗设施，特色餐厅及配备齐全的会议设施,为企业休闲和会务提供理想场所。为了体现对环境保护和企业社会责任，悦榕阁艺品零售店作为与社区合作的窗口，通过经营一系列手工艺品、独特礼品和美容产品，唤醒前来造访的旅行者内心的责任意识。

置身于自然美景之中，悦榕庄酒店都淋漓尽致地展现出当地风土民情，提供一个能够让人完全放松身心的幽雅浪漫空间。本项目将成为广东省第一个悦榕庄品牌酒店。

Banyan Tree Hotel is located at the foot of Mt. Luofu, one of the four most well-known mountains in Guangdong Province. Built by the mountain and facing the lake, the hotel enjoys a picturesque environment. It is one of the serial products completed by Banyan Tree Hotel Group, one of the top operators of boutique resorts, apartment buildings and SPAs.

The Banyan Tree Hotel in Mt. Luofu has about 100 villas, which are categorized into the three major types of Mountain View Villas, Garden View Villas and Waterfront Villas. Each villa has its dedicated entrance. The architectural style of the villas fully embodies the local traditions of Lingnan architecture and the regional characteristics. The building layout takes the traditional villages (fishing villages) in ancient Guangzhou as its prototype and makes the best of the undulating landform and existing vegetation of the site.

The hotel is fully functional with the well-established reception area, meeting area, F&B area, SPA area, guestroom area and sports area. There is also an enterprise village which provides convenient and comfortable accommodations to group visitors. Furthermore, the Banyan Tree SPA, featured restaurants and fully equipped meeting facilities make up an ideal place for companies to organize recreational activities and conferences.

Banyan Tree Hotel fully expresses the local features and customs and provides a peaceful and elegant retreat for people to rest their bodies and minds. This project will become the first Banyan Tree hotel in Guangdong.

项目地点：广东省惠州市
设计时间：2011-2013年
建设时间：2012年至今
用地面积：46万m²
建筑面积：55236m²
建筑层数：2层
建筑高度：14m
合作单位：悦榕庄酒店集团

Location:Huizhou, Guangzhou
Design: 2011-2013
Construction:2012 to date
Site:460,000m²
GFA:55,236m²
Number of floors: 2
Building height:14m
Partner: Banyan Tree Hotel Group

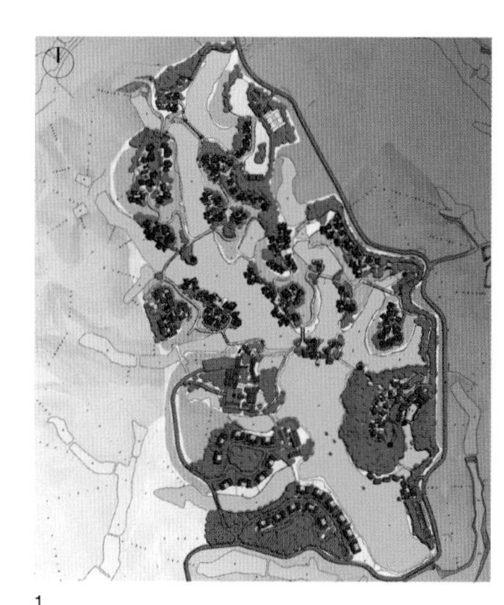

1

2

1　总平面图
Site plan

2　青山脚下，依水而建的度假酒店。水体因标高不同区分为公共服务区域及客房区域等几部分。不同水体间又可相互连通，融为一体
Lying by the water, the resort at the hill foot is divided into several parts including the public service area and guest room area by the waters of varied levels. Those waters are interconnected with each other into one whole system

1

2

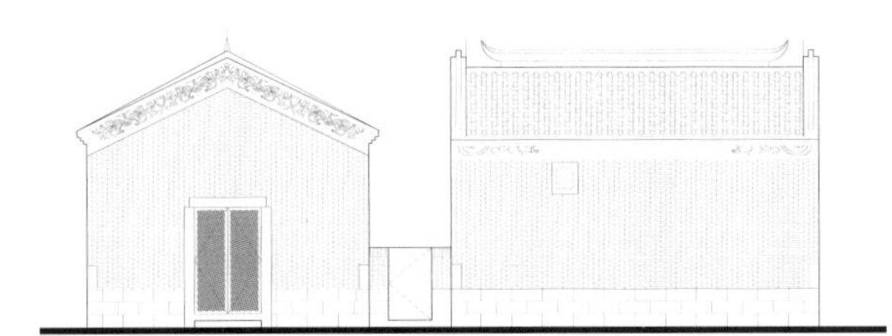

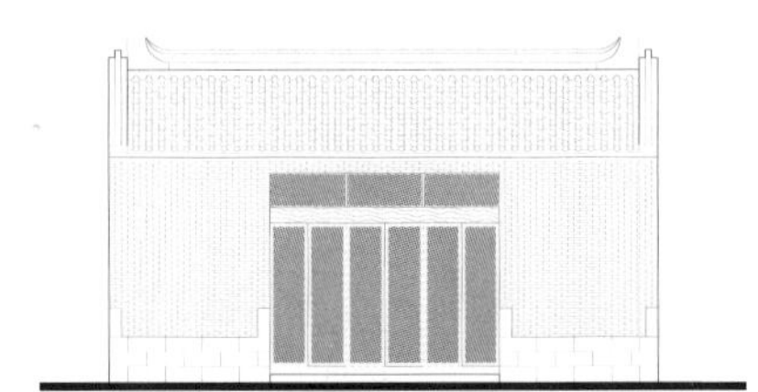

3

4

5

1 别墅群顺流而建尽显水乡风情
Villa Cluster built along the river, fully showing the style of watery towns

2 悦榕庄酒店白云餐厅依水而建，客人可在用餐的同时欣赏秀美景色
The waterfront Baiyun Restaurant offers attractive view to the diners

3 酒店建筑以岭南传统民居的形式为蓝本，立面装饰充满地域特色，采用小青瓦、青砖、砖雕等
The hotel buildings are inspired by the traditional Lingnan architecture. The façades present distinct regional features with small grey tile, grey brick and brick carving

4 酒店客房均为别墅式客房，水岸别墅充分利用临水景观优势，设有室外泡池及亲水平台，住客可乘小艇从亲水平台到达客房
All guest rooms are waterfront villas with outdoor pool and waterfront platform. The guests can reach the guest rooms from the waterfront platform by boat

5 园景别墅效果图
Rendering of Garden Villa

南宁德利·东盟国际文化广场
AICC (ASEAN International Culture Plaza) , Nanning

2010— 广西 南宁 / Nanning Guangxi

东盟德利广场作为南宁凤岭北首个大型办公、商业、生活城市区，集合了文化主题商业街区、商务办公、星级酒店、高尚住宅等全能城市区域服务功能，使本项目区域内生活及商务办公的人可享受到整座城市才能带来的便利。

项目分为南北两个地块，北地块以商业办公为主，面向城市主干道，南地块以居住为主，闹中取静。其用地坡度及高差较大，规划上根据用地原形，使建筑高低相宜，随地形起伏错落有致，巧妙利用坡道、台阶等处理各级高差和消防问题，达到良好的坡地建筑效果。

商业业态的性质决定了其建筑密度必然不小，而建筑密度与绿地率往往是矛盾的两方面。规划中将绿化元素从地面引入各级绿化平台及屋顶，增加绿地率的同时也改善了一般商业综合体生态缺失的不足，成为绿色生态环保综合体。

As the first of its kind in Fenglingbei, Naning City, the project combines all-around urban services and functions such as cultural and themed retail streets, business offices, a starred hotel and high class residential buildings, bringing the various conveniences that only a city is able to offer o those who work and live within.

The project is divided into two plots with one in the north and one in the south. The north plot, facing the city's main road, is mainly for business offices. The south plot enjoys a rather quiet location and is mainly for residences. Due to the gradient and the elevation differences, building heights vary to fit into the undulating terrain in planning. Slopes and terraces are also designed to link up spaces of different elevations and provide accesses to firefighting.

As a result of the trade mix, the building density of the project is inevitably high. In order to handle well the conflict between building density and green area ratio, green elements are also designed on platforms and rooftops of different elevations. It not only helps increase the green area ratio, but also improves the ecological environment that is a common problem among commercial complexes.

项目地点：南宁市青秀区
设计时间：2010—2012年
建设时间：2012年至今
用地面积：67891.64m²
建筑面积：369007.17m²
建筑层数：地上塔楼27层，裙楼4层，地下3 层
建筑高度：99.9m

Location:Qingxiu District, Nanning
Design: 2010-2012
Construction: 2012 to date
Site:67,891.64m²
GFA: 369,007.17m²
Number of floors: 27-floor above-grade tower, 4-floor podium and 3 below-grade floors
Building height: 99.9m

1 运营中的售楼处
Sales Center

2 商业街一角
Shopping street

3 高级公寓
Upscale apartment

4 五星级写字楼
5-star office building

5 整体鸟瞰
Bird's eye view

1

2

3

4

5

三亚凤凰路与迎宾路交界西北侧项目
Project to North-west of Intersection of Feng Huang Lu and Ying Bin Lu, Sanya

2013　海南 三亚市 / Sanya Hainan

项目位于迎宾路与凤凰路交界处，迎宾路往西南指向三亚的滨水城市开放空间，凤凰路则连接了金鸡岭公园与东南面山体，项目区位资源得天独厚，因此本项目有潜力建设成为三亚城市空间节点。

用地周边街区已经建成高密度的居住区，周边开放空间较少，项目用地沿拟建城市广场L形展开。设计的构思是把建筑与城市综合考虑并共同设计，使建筑与城市的开放空间成为有机的整体，营造积极，活跃的城市公共空间环境，从而达到"商业—公共"的双赢结果。

三亚市是亚热带城市，大量种植的椰子树成为三亚城市的文化特征和精神语言，方案中以海边椰树为原型，通过对海风吹过椰子树所形成的特有韵律作抽象化处理，用建筑语言表达在建筑造型之上，使得挺立的建筑犹如轻盈的椰叶，随风起舞。

黎族传统民居是船屋，其屋面酷似倒过来的木船，十分适应当地气候。因此我们在建筑造型中加入了船屋的抽象元素，运用了适合三亚气候的大飘檐设计。

整体建筑通过前面零散的建筑体量，结合前后广场的联合设计处理，使得建筑犹如身处一个大型城市公园当中，城市广场通过活泼的放射线创造出灵活的空间，使得建筑在拥有浓厚的商业氛围的同时也成为三亚这座城市的市民公共空间重要节点。

The Project is prominently located at the intersection of Ying Bin Lu and Feng Huang Lu. The former leads to the riverfront urban open space toward the southwest while the latter connects Jinjiling Park and the massif in the southeast. Therefore, the project has the potential to become a key node within the urban space.

The Project extends in L shape along the planned urban square. The design intends to develop the buildings and the open spaces of the city as an organic whole to foster an environment of proactive and dynamic urban public space, thus achieve a win-win situation for "retails – public spaces".

Referencing the coconut trees, the cultural icon and spiritual vocabulary of the city, the building form features the abstraction of the unique rhythm created by the gentle sea breeze blowing over the coconut palms, creating towering yet light-footed buildings.

The abstract element of boat house, the traditional housing of Li people, as well as large floating eaves are incorporated into the architectural form to cope with local climate

With scattered building volumes in the front, and integrated treatment of the front and rear squares, the building seems to be situated inside a large urban park. The lively radial lines help create flexible spaces on the urban squares, fostering strong commercial atmosphere while establishing a key node of civic public spaces in Sanya.

项目地点：海南省 三亚市
设计时间：2013年
用地面积：17101m²
建筑面积：76960m²
建筑层数：地上40层，地下1 层
建筑高度：145m

Location:Sanya, Hainan Province
Design:2013
Site:17,101m²
GFA: 76,960m²
Number of floors: 40 above-grade floors and 1 below-grade floor
Building height: 145m

1　总平面
Site plan

2　设计演变过程
Design evolution

3　把建筑与城市综合考虑并共同设计，使建筑与城市的开放空间成为有机的整体，营造积极，活跃的城市公共空间环境，从而达到"商业—公共"的双赢结果
The architecture and the open spaces of the city are considered and designed as an organic whole to foster proactive and dynamic urban public space, thus achieve a win-win situation for the "retails - public space"

4-5　建筑效果图
Building rendering

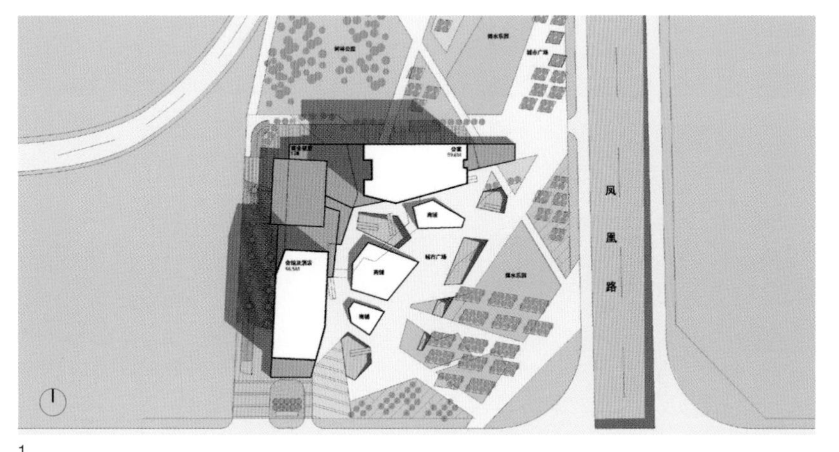

1

2

3

4

5

广州科学城科技人员公寓
Scientists' Apartment, Guangzhou Science City

2006-2010　广州 萝岗 / Luogang Guangzhou

广州萝岗新城是广深经济走廊上的科研孵化中心、广州东部地区的现代化服务中心。为吸引海外留学人员归国创业，满足在广州萝岗新城工作的外籍人士、留学归国人员及科技人员、专家学者等住宿的要求，广州萝岗区启动了广州科学城科技人员公寓的建设项目。

建筑采用围合式布局，两栋塔楼公寓和两栋板式多层公寓围合出内部庭院。庭院内布置有会所建筑与泳池、下沉花园等公共活动空间。科技人员公寓的定位、服务对象及标准多样化，住宅单元类型多。在保证居住空间私密性的前提下，设置促进邻里交往的公共空中花园。通透的空中花园、与公寓绿化阳台、自由组合的遮阳百叶构成了独特的、富有灵气的建筑形象。两栋塔楼采用微纺锤形的剖面设计，成为当地的标志建筑。平面周边的柱子上下端采用斜柱，中段直柱连接的形式，并采用直线预应力钢筋解决转折处楼面构件受拉问题，避免建筑结构的竖向不规则的不利影响。

科技人员公寓主要采用了被动式太阳能设计。居住空间坐北向南，间距适中，适应岭南地区的气候特点；架空层和空中花园能够形成顺畅的风之通路；绿化露台与竖向百叶相互作用，达到有效的隔热效果，保证舒适的室内环境。另外，项目还使用了大量能减少对环境带来负荷的新技术、新设备，以实现绿化环境的目标。如雨水回收系统、冷交换热泵供水系统、节水洁具、智能电气系统、智能多联体空调系统、LED节能园林灯等，有效的降低运营费用。

To attract the returned overseas students to start business and address the accommodation of expatriates, returned overseas students, S&T Professionals and experts working in Luogang New Town, Luogang District government launched the Project. Our design team was awarded with the design contract after winning the international design competition.

The inner courtyard framed by two tower apartments and two slab-type multi-floor apartments is provided. The apartment is featured with diversified positioning, service subjects, standards and apartment types. With ensured privacy, multiple public sky gardens are provided to enhance the sociality. The two towers with fusiform section become the representative buildings. The linear pre-stressed rebars used of building components at the turning avoid the adverse impact brought by vertical irregular structure.

With passive solar design, the residential spaces face the south with appropriate interval to respond to the local climate features. In addition, multiple new technologies and equipment have been applied to greatly cut the environmental loads and operation cost.

项目地点：广州市萝岗区
设计时间：2006—2007年
建设时间：2008—2010年
用地面积：39957m²
建筑面积：105358m²
建筑层数：22层
建筑高度：77.75m（塔楼），31.3m（板式公寓）
合作单位：日本佐藤综合计画
曾获奖项：2006年国际竞赛第一名
2011年第六届中国建筑学会建筑创作优秀奖
2011年广东省注册建筑师优秀建筑创作奖
2011年度广东省优秀工程二等奖

Location: Luogang District, Guangzhou
Design: 2006-2007
Construction: 2008-2010
Site:39,957m²
GFA: 105,358m²
Number of floors: 1 below-grade floor, 22 (tower) / 9 (slab-type building)/ 2 (clubhouse) above-grade floors
Building height: 77.75m (tower), 31.3m (slab-type building), 10.55m (clubhouse)
Partner: AXS
Awards:
The First Place of national competition(2006)
Excellent Award of the 6th ASC Architectural Creation Award (2011)
Excellent Architecture Creation Award of Guangdong Chapter of Association of Chinese Registered Architects (2011)
The Second Prize for Excellent Engineering Design Award of Guangdong Province (2011)

1　总平面图
Site plan

2　纺锤形双塔成为萝岗区标志性建筑
The spindly twin towers are made the landmarks in Luogang District

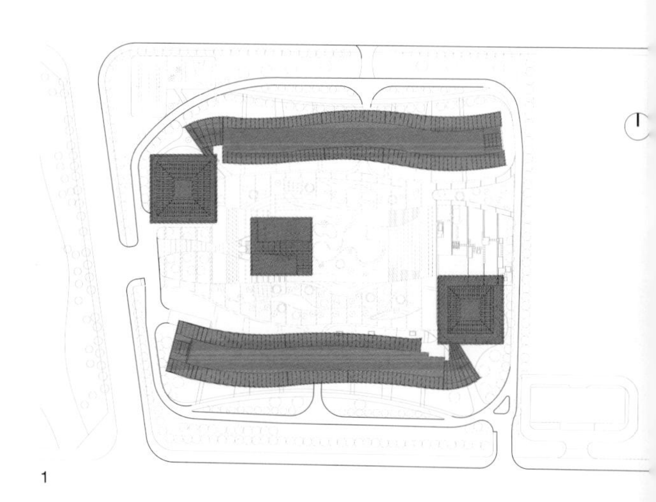

1

2

1

1 内部庭院，庭院内布置有会所建筑与泳池、下沉花园等公共活动空间
Internal courtyard with club house, swimming pool, sunken garden and other public activity spaces

2 住区主入口景观
A view of the main entrance to the compound

3 剖面图
Section

2

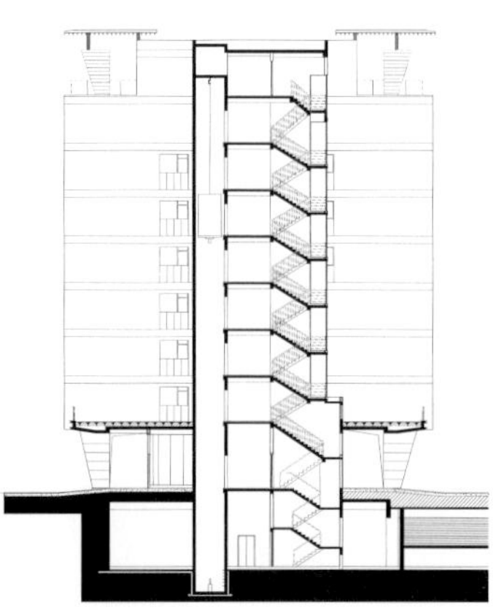

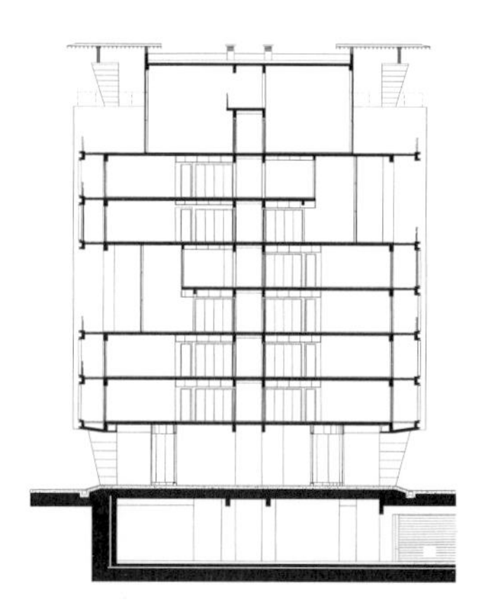

3

4

5

6

4 位于庭院中央的会所是一座两层的小建筑，主要功能包括超市、健身室、咖啡厅等
The club in the center of the courtyard is a small two-storied building accommodating functions including supermarket, fitness room, coffee house, etc.

5 板式公寓：巧妙利用板式公寓空中花园两侧的山墙面，采用鲜艳色彩的外墙涂料，赋予每一个住宅组团独特的个性
Slab-type apartment: Expertly use the gable wall on both sides of the sky garden of the slab-type apartment, which is coated in vivid color, giving unique individualities to each residential cluster

6 建筑立面细节：垂直百叶、屋面遮阳百叶及大进深露台相互作用，保证整个建筑的节能效果
Façade details: Vertical shutters, rooftop sunshading shutters and the large-depth terrace have interaction on each other to ensure the energy-efficiency effects of the entire building

1

2

3

1　在保证居住空间私密性的前提下，设置促进邻里交往的公共空中花园。通透的空中花园、与公寓绿化阳台、自由组合的遮阳百叶构成了独特的、富有灵气的建筑形象
While assuring the privacy of the residential space, provide public sky garden to encourage neighborhoods interaction. The transparent sky garden, apartment greening balcony and free-combined sunshading shutters form the distinctive and vivid architectural image

2-3　会所中庭空间
The atrium of the club house

4　多层板式公寓利用顶层的钢架与高层塔楼铰接相连，以增强结构的稳定性。板式公寓末端采用V形钢管混凝土斜柱支撑从下至上逐渐增大的绿化平台
Multi-storied slab-type apartments are hinged with the top steel structure and the high-rise tower, so as to increase the stability of the structure. At the end of the slab-type apartment, V-shape steel-pipe concrete tilted-columns support the green platform that expands in size along with the height

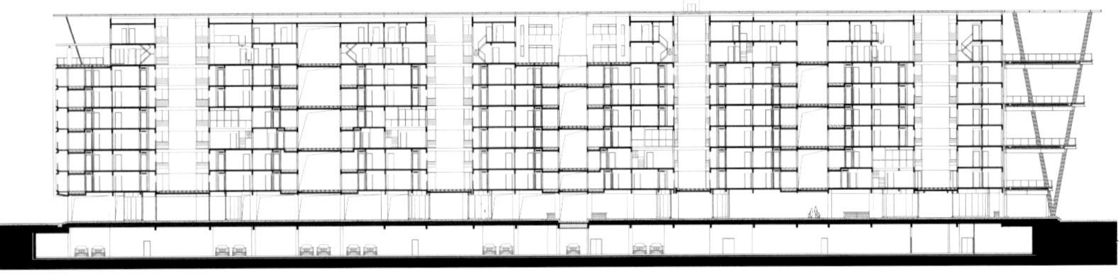

4

富现代科技气息的内庭园景
Inner courtyard view with modern scientific and technological atmosphere

香港新福港地产 · 广州萝岗鼎峰
SFK·DF Project, Luogang Guangzhou

2011－ 广州 萝岗 / Luogang Guangzhou

2

广州市从亚运会前夕启动了全市范围的三旧改造进程，旨在更有效、更合理的利用有限的城市土地，创造更好的城市生活环境。本项目性质也属于此类。项目选址位于萝岗区政府附近，北侧为善坑山，南侧为城市主干道，面向广州演艺中心及拟建的商业综合体，附近是地铁14号线萝岗站，地理位置十分优越。“因地制宜，筑半山好住宅；融会贯通，造岭南新社区”是本项目设计的重要宗旨。

项目包含回迁区及融资区两部分。回迁区位于线坑村原址上，是线坑村民的回迁住宅及出租物业。在布局上，充分考虑了当地的风俗生活，布置了满足祭祀庆典的祠堂广场，及各种小尺度的邻里交往空间，使村落的习俗文化得以延续。融资区位于善坑山山腰，是面向房产市场的中高档商品住宅。设计充分利用山地地形，创造高低错落，空间体验丰富的住区环境；建筑布局疏密有致，尊重城市轴线，延续城市视廊，使山景融入社区及城市景观中。

整体建筑造型简洁现代，体现时代特色。住宅顶层采用悬挑的百叶屋架，既起到屋面隔热的作用，又能体现岭南建筑轻盈的神韵。两区的建筑造型整体色调一致，细部上有所区别，平衡了投资造价、住区氛围营造等多个方面因素。

Right before the Asian Games 2010 Guangzhou, the city launched the Three-olds Redevelopment. This project falls within this category. The site is close to Luogang Station of the Metro Line 14. The design philosophy is "to build good hillside residences in view of the local conditions and develop a new Lingnan community by integration of traditional architecture".

The project comprises the resettlement housing area and the developer-financed residential development. The former is located at the original place of Xiankeng Village. The developer-financed residential development located on the hillside of Shankeng Hill is to develop the market-oriented medium- and high-class commercial residences. The design makes best use of the hilly terrain to create an elaborately devised architectural layout, showing respect for the city axis, extending the visual corridor and bringing the mountain view into the community and the city.

The building forms are in general, simple and modern, reflecting the features of the times. The top floors of buildings in the developer-financed residential development are installed with cantilever louver roofs, which not only serves the purpose of insulation but also reflect the ethereal grace of Lingnan architecture. The buildings in both areas generally share the same color tone with certain differentiated details.

项目地点：广州 萝岗中心区
设计时间：2011—2014年
建设时间：2011年至今
用地面积：12万m²
建筑面积：约42万m²
建筑层数：1层—28层
建筑高度：99.9m

Location: Luogang Central Area, Guangzhou
Design:2011 2014
Construction:2011 to date
Site:120,000m²
GFA:about 420,000 m²
Number of floors:1-28
Building height:99.9m

1 总平面图
Site plan

2 项目建设中。会所钢琴般的形态与泳池、花园融合一体。前期用作项目的销售中心
Project under construction. The piano-like club house is perfectly integrated with the swimming pool and garden, and will be used as the sales center of the Project at the initial phase

3 项目建设中
Under construction

4 鸟瞰效果图
Bird's eye view

5 回迁区住宅效果图
Rendering of the Resettlement Housing Area

6-7 低点夜景效果图
Night view rendering from low viewing point

4

1

3

5

6

7

佛冈汤塘镇鹤鸣洲温泉度假村
He Ming Zhou Hot Spring Resort

2012-2013　清远 佛冈 / Fogang Qingyuan

黄花湖风景区拥有天然温泉资源，以及开发度低、人口密度低等区域特点，为打造尊贵舒适的温泉度假区提供了非常优厚的自然景观条件，具有良好的发展潜力。规划总用地面积约73万m²，其中包括：地块A（规划建设用地面积约为27525m²）和地块B（规划建设用地面积约为32007m²）。地块A和地块B之间为黄花河，南面有优美的小湖，其余三面为自然生态山景，自然景观秀丽。

建筑设计追求简洁、明快、生态的现代建筑风格。地块A为高层建筑，立面强调垂直感，细化建筑体块的同时运用随机旋转、错位的方式塑造出有机的建筑形态，整体上呼应了地形地貌。地块B为低层建筑，重点考虑立面的水平感与屋顶形式。利用半围合式的建筑平面，使各个空间能够得到自然通风采光，也使建筑体量既多元化又富有色彩，但同时又保持了建筑的整体性。首层采用大面积落地门窗，使室内外空间融为一体，在室内能欣赏到泳池与私家花园景观。二层主人房与其他卧室也都设有大面积落地窗与休闲阳台。除坡屋顶深檐口有很好的遮阳效果外，其他主要门窗也设有百叶式遮阳板。建筑设计结合佛冈县的气候条件，注重自然通风强调生态自然的特点。

立面细部处理上采用大量的仿木构件，采用亚热带休闲度假风格，强调建筑的通透性和通风效果,同时把一些热带元素以现代方法加以表现，令建筑具有强烈的标志性，也充分营造了一体化的休闲度假氛围。

Surrounded by hills on the other three directions, Huanghua Lake enjoys a nice natural environment with hot springs, low development and population density, hence an ideal location for upscale spring resort development. This project is planned with a site area of about 730,000m². Huanghua River runs between Plot A and Plot B and forms a small beautiful lake in the south of the site.

The architectural design strives to create modern buildings that are concise, lively and eco-logical. Plot A is mainly for high-rise buildings with emphasis put on the verticality of façade. When designing building blocks, approaches such as random rotating and staggering are adopted to create organic building forms. Plot B is mainly for low-rise buildings that emphasize the horizontality of façade and form of roof. Semi-enclosed plans are used to ensure that each space of the buildings is well lit and ventilated. The plans not only enrich the types of building volume but also maintain the completeness of the buildings. Large full-height glazed doors and windows are installed on ground floor so that guests can enjoy the views of the private pool and the private garden from inside the room. F2 is also installed with large full-height glazed windows with terraces. In view of the climate conditions of Fogang County, the architectural design attaches importance to natural ventilation and highlight the features of ecology and nature..

Synthetic wooden members are used on façade to bestow the buildings a subtropical appearance and deliver a resort area that integrates multiple uses.

项目地点：广东省清远市佛冈县黄花湖风景区
设计时间：2012—2013年
建设时间：2013至今
用地面积：73840.8m²
建筑面积：108095.1m²
建筑层数：2-22层
建筑高度：6.8-61m
合作单位：Aedas凯达环球国际设计公司

Location: Fogang County, Qingyuan, Guangdong Province
Design:2012-2013
Construction: 2013 to date
Site:73,840.8m²
GFA: 108,095.1m²
Number of floors: 2-22
Building height: 6.8-61m
Partner:Aedas

1　总平面图。充分利用地块资源，力求实现户户亲水。保证每户内部的私密性和组团之间的开放性，并能便利享受周边资源
Master plan. Site assets are fully employed to maximize the riverfront space to every house, ensure privacy for the interior spaces of each house and the openness between various clusters, and offer easy access to peripheral resources

2　高层造型灵感来源于温泉中使用的木桶，叠加后运用水波形态细化建筑体量
The form of the high-rise is inspired by the wooden casks used in the hot spring, which are superimposed and then detailed into the pattern of ripples

3　整个项目天际线波幅平缓，高层公寓与别墅高度谐和，相得益彰。高层建筑疏落有致，保证每户都享受到周边优美的景观
The project presents gently wavy skyline with harmonious height contrast between the high-rise apartments and the villas. The high-rise buildings are staggered in layout to ensure nice view for every apartment

4　采用副热带休闲度假风格，强调建筑的通透性和通风效果。别墅内除建筑主体及私家花园外，还包含有度假风情的亲水泳池及温泉SPA池。采用落地玻璃门窗，将自然美景引入室内
Subtropical resort style is introduced in the design to emphasize the transparency and ventilation effect of the building. In addition to the main building and private garden, resort-style swimming pool and hot spring SPA pool are also provided in the villa. Floor-to-ceiling glass doors and windows help introduce the natural scenery into the interior

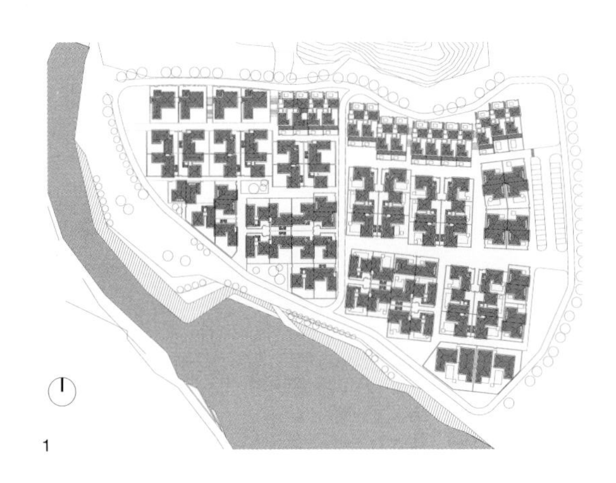

1

2

3

4

香港新福港地产 · 鹤山峻廷湾二期
SFK·Jun Ting Wan (Phase II), Heshan

2012— 江门 鹤山 / Heshan Jiangmen

鹤山峻廷湾小区是由新福港地产开发建设的大型优质住宅小区，项目与中心城区一桥之隔，可以充分共享城区的学校、医院、商业街等各项配套。项目定位为"城市生活典范"，是鹤山市场上超大规模和高档次的居住社区；使消费者能够充分感受到城央高档社区与郊区社区的区别，现代、时尚、简约、大气的设计使小区呈现出尊贵感，营造出城市中央高档社区生活的典范。

建筑整体布局充分利用项目周边面向河流、鱼塘、田地等自然景观资源，并营造中央人工湖，创造内部景观提升产品素质。住宅群体布置上避免建筑物之间的相互遮挡，满足高级住宅对日照、间距、自然采光、自然通风的要求，并充分考虑景观利用，研究前后排间高度与间距关系，巧妙利用毗邻单元间的错位，令各户取得观景资源平衡。建筑在统一的总体风格中寻求形式的变化多样，以空间开合、高低错落、形态对比等形成丰富的空间层次，同时强调资源占有的均好性，为某些条件不佳的单元提供其他方面的合理补偿，强调朝向、日照、通风均好性；强调景观空间的变化及功能的合理利用，为各种户型创造相应卖点；强调景观的均好性，景观资源配置时遵循档次高的物业配好景观的原则。

建筑立面设计强调功能化，避免无功能的过度装饰；采用现代建筑风格，通过运用面砖、涂料、石材、铝合金板等装饰材料演绎现代、时尚、简约、大气的立面形象。合理组织窗位及室外空调机百叶位置，使其成为立面效果的积极元素，使立面效果更为整洁、大方。建筑户型创新的同时，尽可能完善普通户型结构，做到方正、实用、通风采光好、动静分区好、空间组合合理、空间利用率高，做到每户向南，同时采用分列电梯的布置使住户有更好的私密性。

Jun Ting Wan is a huge quality residence community developed by SFK, and the project is a bridge away from downtown, with full access to the schools, hospitals and business streets among other supporting facilities in downtown. The project, branded as "a model of urban living", is a residence community of supersize and top notch in Heshan property market.

The general layout of the project takes full advantages of natural landscapes surrounding the site including river, fish pond and field; moreover an artificial lake is built to improve the class through inner landscape. The arrangement of building blocks meets the requirements of high-grade residence towards sunshine, spacing, natural lighting and ventilation. The building seeks changeable forms in the uniform style to create layered spaces through the opening and closing of space, changed heights and patterns contrast. It pays attention to the change of space and functions utilization to create merits for each type of unit, and to the even distribution of landscapes following the principle of landscapes corresponding to each one's grade.

The façade design emphasizes on functions. It adopts modern architectural style, and expresses the façade of modern, stylish, simple and grand nature through usage of decorating materials including bricks, painting, stones and aluminum alloy plates. Apart from innovation in unit types, normal unit setup is to be improved and perfected to ensure its suitability, good ventilation and lighting, well partition of the movable and the still, reasonable space combination, high utilization of space as well as south-facing of each unit. Also, separated arrangement of elevators is adopted to give residents more privacy.

项目地点：广东省江门市鹤山
设计时间：2012—2013年
建设时间：2013年至今
用地面积：6.15万m²
建筑面积：25.95万m²
建筑层数：地上23—32层，地下1层
建筑高度：99.4m
合作单位：香港嘉柏建筑师事务所

Location: Heshan, Jiangmen City, Guangdong Province
Design:2012-2013
Construction: 2013 to date
Site:61,500m²
GFA: 259,500m²
Number of floors: 23-32 above-grade floors and 1 below-grade floor
Building height: 99.4m
Partner: Gravity Partnership (HK)

1 鸟瞰图
Bird's eye view

2 总平面图
Site plan

3-5 小区低点透视图
Low view-angle community perspective

1

2

3

4

5

广西扶绥金源财富广场
Jinyuan Fortune Center, Fusui Guangxi

2011－　广西 崇左市 / Chongzuo Guangxi

项目选址位于扶绥县新城的城市中心，毗邻城南商业集中区和政府行政区，地理位置十分优越。金源财富广场包括住宅、办公、酒店式公寓、大型商场及配套公建设施。项目用地面积8.7hm²，其中包括：西地块（规划建设用地面积56572m²）和东地块（规划建设用地面积30768.67m²）。根据西地块与城市广场接壤的特点，考虑西侧地块主要作为集中商业区，东地块形成独立住宅区。

建筑布局南低北高，错落有致，主要为东南及南北朝向，互不遮挡。根据用地、视线、景观等因素考虑，部分住宅朝向略作偏斜，保证最大程度利用小区中心花园景观资源，并营造出富有变化的园区空间，酒店式公寓塔楼错落布置，互不干扰，均有良好视线，且空间形态上形成整体韵律感。项目充分考虑景观设计，绿化庭院渗透到建筑物的周边，围绕中心绿化、水景、建筑小品，景观与建筑浑然一体，使建筑空间和非建筑空间相辅相成，形成良好优雅的内部环境。

建筑立面采取规则、简洁、明快的设计手法，避免缺乏功能意义的过度装饰。通过运用不同颜色的面砖、涂料、石材、铝合金百叶等装饰材料演绎现代建筑时尚、大气，体现时代特色。大型商场设计色调上着重表现出浓厚的商业气氛，住宅用色则采用温馨的暖色调，与绿色植物相互辉映，烘托出温馨的生活气息。

The project is prominently located in the center of Fusui New Town and in the vicinity of the commercial district and the administrative area, comprising residential buildings, offices, serviced apartments, a shopping mall and supporting public facilities. The site covers a land area of 8.7 hectares, including West Plot (56,572m²) for commerce and East Plot (30,768.67m²) for residence.

Buildings in the north are taller than those in the south and most buildings are southeast/ south-oriented. Depending on the site conditions, views and landscape, the orientation of some residential buildings will be slightly deviated to maximize the views to the central garden and create diverse garden spaces. The serviced apartment towers are carefully located to ensure quality sightlines and avoid interfering with each other. As for the landscaping, the green courtyards are designed around the buildings which elegantly interweave with the central green space, waterscape and site furnishings. This way, the building spaces and non-building spaces are complementary to each other to form an attractive internal environment within the project.

The building façades are regular, concise and lively without those excessive non-functional decorative elements. Instead, decorative materials of different colors, such as face bricks, paints, stones, aluminum alloy louvers, are used to present the trendy and elegant quality of the modern buildings. In the design of the large shopping mall, the selected color tones emphasize the strong commercial atmosphere of the space. Meanwhile, warm color tones on residential buildings respond to and complement with the green plants to create a cozy living environment.

项目地点：广西崇左市扶绥县
设计时间：2011年至今
用地面积：87340.67m²
建筑面积：430203m²
建筑层数：地上31层，地下2 层
建筑高度：99m

Location:Fusui County, Congzuo City, Guangxi Province
Design: 2011 to date
Site:87,340.67m²
GFA: 430,203m²
Number of floors: 31 above-grade floors, 2 below-grade floors
Building height: 99m

1　总平面
Site plan

2　项目由住宅、办公、酒店式公寓、大型商场及配套公建设施组成
The project is comprised of residence, offices, serviced apartments, a large shopping mall and supporting public facilities

3　鸟瞰图
Bird's eye view

4　夜幕降临，时尚、现代的大型商场，在霓虹灯的映衬下，闪烁夺目
When night falls, the trendy and modern large mall dazzles in the neon lights

5　住宅用色则采用温馨的暖色调，与绿色植物相互辉映，烘托出温馨的生活气息
Warm color tones on residential buildings respond to and complement with the green plants to create a cozy living environment

2

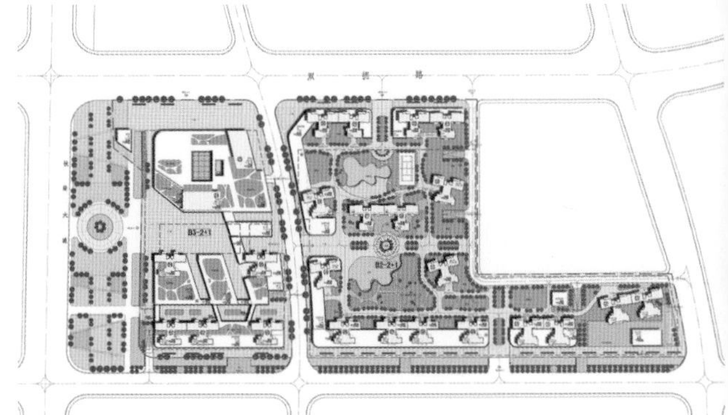

1

3

4

5

从化新城市民之家
Citizens' Home of Conghua New Town

2013　广州 从化 / Conghua Guangzhou

在过去的几年里，广州市实施了"北优"战略，推动了以从化市为核心的北部片区的蓬勃发展。从化将成为广州市北部的城市副中心，由此带来城市"提质扩容"，从化新城的建设顺理成章，而新城急需一个集中为市民政务服务的"平台"，本项目在这样的背景下应运而生。

位于新城中轴线和河湖的北岸，从化新城市民之家是新城的核心公共建筑，庞大而多样的功能使得它本身就仿佛一个微缩的城市。设计提案将其整合为一个独立的城市体块，一个都市平台。都市平台连接着会议展览、综合审核、公共服务三大功能区，并将基础设施及公共服务配套等后场和支援空间合并成单一而有效率的整体。平台上设置连续的中心花园，同时向南北延伸，成为"H"形城市花园，有机连接城市空间与市民之家，并与其他平台花园一起成为市民之家中立体的、最具活力的城市空间。在这里，多样化的使用对象及活动内容都可以在同一时空发生和交织，进而叠加并产生新的绵绵不断的城市活力。行人可沿各分支步道向中心聚集，到达位于市民中心南面首层宽敞开放的、可以容纳大型活动的公共空间。这里也将成为地块的主要广场和城市的网络中心。公众也可以通过室外步道去到展览中心，了解当代最前沿各种资讯。一条立体环形流线将引导观众进入各种展厅和观展体验，流线自然延伸到屋顶花园，在此可获得纵观新城景观清晰而全面的参观感受。

市民之家是一个标志性建筑，也是新的城市肌理的一个有机部分，它带给我们的不仅仅是标志性，还有亲切的社区和城市活力，以及它们交织叠加而成的连续都市生态景观。在市民之家这样新型的都市综合体中，高效的都市营运和休闲多样化的都市生活在这样新型的城市空间中合而为一。更多公众的自由参与使得市民之家不再是一个城市孤岛，而是具有多方连接的城市多面体，连续而开放的城市空间，犹如打通的都市经脉，最终为我们带来的，不只是短暂的灿烂，而是长久的活力。

With the implementation of the "Optimize the North" strategy by Guangzhou Municipality, the northern part of the city, with Conghua at the core, has enjoyed a vigorous development over the past few years. The New City requires a centralized platform to provide government services. This Project emerges in response to this need.

Citizens' Home of Conghua New Town is a large core public building with diverse functions, which makes it a miniature city. The design proposal tries to integrate the city as an independent urban block and an urban platform, connected with the three major functional areas for convention and exhibition, comprehensive examination and approval, and public service respectively, while the infrastructure, public service supporting facilities and other backcourt and supporting spaces will be combined into a single and efficient whole. On the platform, a continuous central garden will be built, which will extend to the north and south, forming an H-shape urban garden with organic connection to the urban space and Citizens' Home. The central garden, together with other gardens on the platform, will constitute a three-dimensional, most vibrant urban space. A three-dimensional ring circulation will guide visitors into different exhibition halls. And the circulation will naturally extend to the roof garden, where people will be offered a clear and paramount view of the landscape of the New City.

The Citizens' Home is a landmark building, and an integral part of the new urban fabric. In a new urban complex like the Citizens' Home, efficient operation and diversified urban life become one in the new urban space. The involvement of more people will make the Citizens' Home no longer an isolated urban island, but a well-connected urban polyhedron and a continuous open urban space. Eventually, it will bring us not just transient brilliance, but also a long-term vitality.

项目地点：广州 从化
设计时间：2013年
用地面积：18.7万m²
建筑面积：24万m²
建筑层数：7层
建筑高度：28m
曾获奖项：2013年 国内竞赛第一名

Location: Conghua, Guangzhou
Design:2013
Site:187,000m²
GFA: 240,000m²
Number of floors: 7 floors
Building height: 28m
Awards: The First Place of national competition(2013)

1

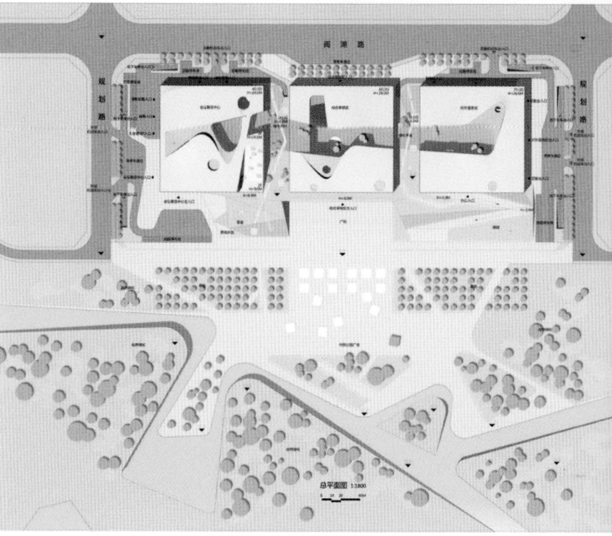

2

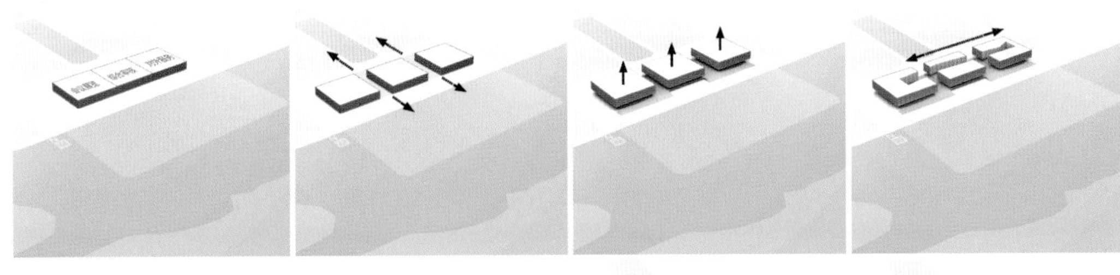

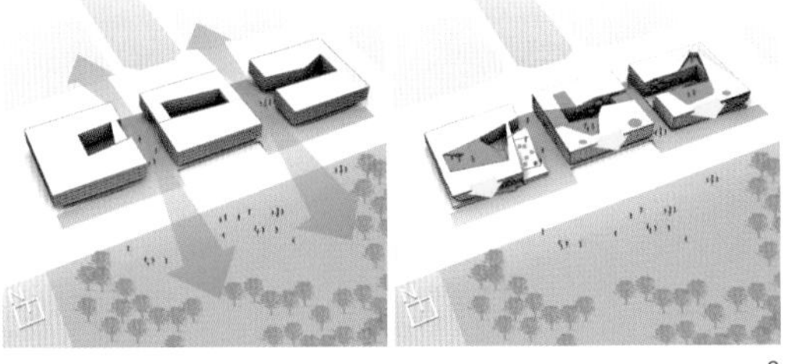

3

1 鸟瞰图：三个干净的体量嵌在新城中心，简约的体量蕴含了无限变化的内部空间
Bird's eye view: Three clear-cut volumes containing varied interior spaces are embedded into the center of the New City

2 总平面图
Site plan

3 整合、打开、提升、串联、引入、丰盈，一步一步生成从化新城的市民之家
The spaces are integrated, unfolded, upgraded, interconnected, introduced and enriched to gradually create the Citizens' Home in Conghua New Town.

1

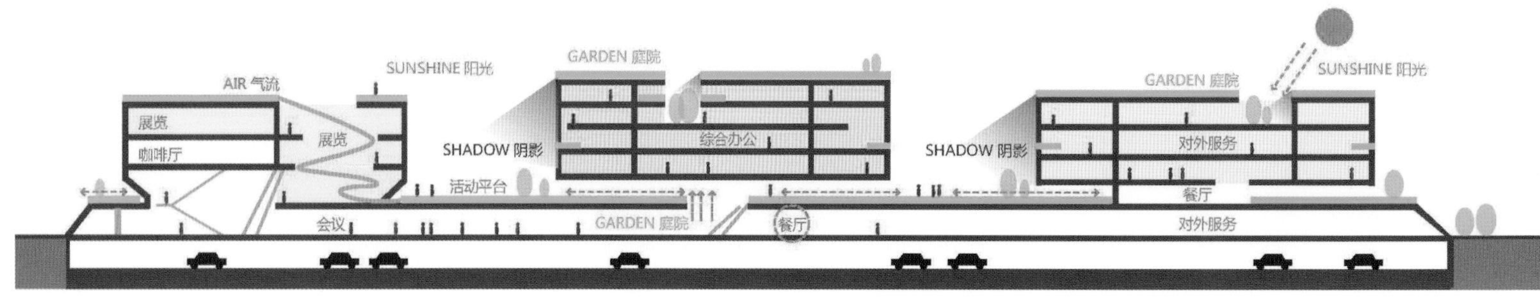
GARDEN 庭院
SUNSHINE 阳光
AIR 气流
展览
咖啡厅
展览
SHADOW 阴影
综合办公
GARDEN 庭院
SUNSHINE 阳光
SHADOW 阴影
对外服务
活动平台
餐厅
会议
GARDEN 庭院
餐厅
对外服务

2

3

1 绿色平台沿H形花园延伸，直至城市广场，建筑和城市空间在此融为一体，形成连续的岭南都市生态景观
Green platform extends along H-shape garden till the urban square where the building and the urban space are integrated to form the continuous and ecological urban scene of Lingnan

2 与地形协调一致的绿化平台将三个体量串在一起，在空间上联系了三个不同使用功能。剖面上深入考虑了生态，竖向交通整合的问题
The terrain-oriented greening platform link up the three volumes and spatially connects the corresponding functions. The section incorporates considerations on ecological issues and vertical

3 市民中心夜景效果图
Night view of Citizens' Home

夜幕降临，市民之家如同三颗钻石闪耀在新城之央。
The Citizens' Home glitters in the center of the New City when night falls.

广州气象卫星地面B站区业务楼
Business Building of Guangzhou Meteorological Satellite Ground Station Zone B

2009-2013　广州 萝岗 / Luogang Guangzhou

项目地点：广州 萝岗
设计时间：2009—2010年
建设时间：2010—2013年
建筑面积：5539m²
建筑层数：地上三层，地下一层
建筑高度：13m
曾获奖项：广东省注册建筑师协会优秀建筑创作奖

Location: Luogang, Guangzhou
Design:2009-2010
Construction: 2010-2013
GFA: 5,539m²
Number of floors: 3 above-grade floors and 1 below-grade floor
Building height: 13m
Awards: Excellent Architecture Creation Award by Guangdong Chapter of Association of Chinese Registered Architects

建设方一直使用的气象卫星地面站A站区处于广州市区内，并不适合气象卫星业务发展需求。早在多年前国家已计划在广州郊区另行建设B站区，通过多种因素比对确定了项目所在地。到目前为止，该站是全国仅有的四个气象卫星地面站之一，也是华南地区仅有的气象卫星地面站。项目包含了一系列建构筑物，其中业务楼除担任主要的技术任务外，局部区域还兼参观教育功能，因此业务楼在外形上最接近一般认知范围的建筑物形象。

众所周知，气象变化具有高度的复杂性与不确定性。而气象卫星业务中重要的一环是接收来自太空中气象卫星所拍的大气云图。业务楼建筑造型上并没有代入气象云图的具体形象，而是发掘其内在的复杂性与不确定性，以具备相当视觉冲击力的棱角和线条隐喻大自然气候的威力，建筑独特的外形成为B站区展示其工作复杂性和不确定性的最好代表。在强大的自然力量面前人类要获得存在空间，必须协调与自然的关系，正如在如此独特的建筑外形下内部空间如何保证满足各项功能一样。设计初期整合了所有条件，包括建设要求、使用要求、设施安装要求、施工工艺要求等，尽可能在不能正常使用的空间内化解形体和结构的矛盾，以及安排设施或设备。

业务楼从协调和利用自然山风出发，结合调查岭南地区多年的生活经验，大多数公共空间并不安装空调系统，通过空间组织设计在春夏秋均可通过自然通风的方式达到一定舒适度。同时考虑到冬季山风较冷的情况，多数主要活动空间均进行了节能设计。事实上在业务楼并未完全装修完成的情况下，使用单位对局部区域舒适度的反映良好。

Guangzhou Meteorological Satellite Ground Station Zone A so far is one of the only four meteorological satellite ground stations nationwide, and the only meteorological satellite ground station in southern China. The Project includes a series of structures, among which the Business Building is mainly used to serve the needs of performing technical tasks, and particular areas will also be used for educational purpose.

The Business Building tries to explore its inherent complexity and uncertainty in order to create corners and lines with fairly strong visual impact, which metaphorically represents the power of nature. Its unique architectural shape can best reflect the complexity and uncertainty of the work being undertaken in Station Zone B. The design initially integrated all conditions, including requirements for construction, usage, facility installation and construction technology, trying, as far as possible, to resolve the conflicts between form and structure in spaces that cannot be used normally, and to arrange facilities or equipment.

Based on the coordination with and utilization of mountain wind, and with appropriate spatial design and arrangement, the comfort of the Business Building can be achieved to a certain degree through natural ventilation in spring and autumn. Most of the main activity spaces are provided with thermal insulation structures. In fact, even when the decorations of the Business Building were only partially finished, we had already received positive comments from the users about the comfort of particular areas.

1　总平面图
Site plan

2　鲜明的建筑体块形成交叠错落的空间
Distinctive building massing creates interestingly staggered spaces

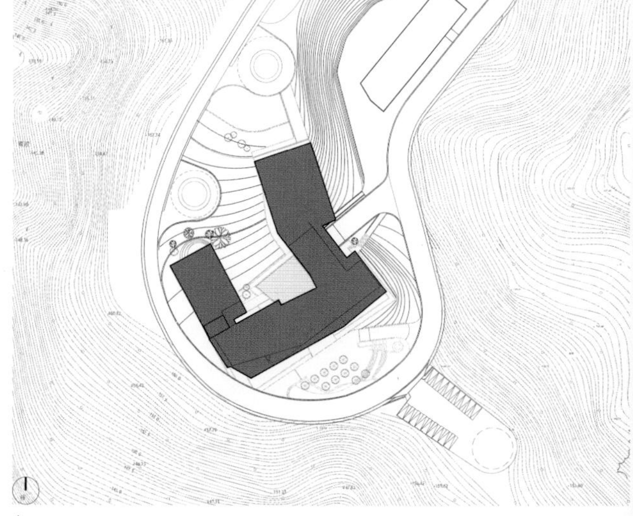

1

1

2

1 清晨的建筑外玻璃映照着东方鱼肚白
External glazing of the building reflects the grey dawn twilight

2 从东南方向看建筑物，建筑与山体完美契合
Perfect integration of architecture and mountain viewed from southeast

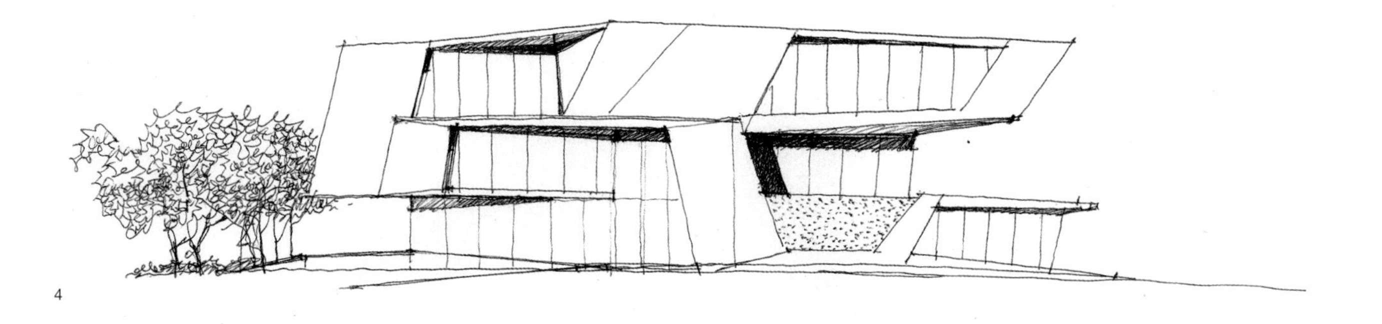

4

3

3 掩映在植物间的建筑一角
Part of the building set off among the plants

4 设计手稿
Design Sketchs

5 雕塑感强烈的建筑立面局部
Sculpture-like façade

6 建筑师有意控制西立面的开窗面积，有效降低了西晒对室内的影响
The fenestration area in the west façade is carefully controlled to reduce exposure of the interior to the sunlight in the west

5

6

1

2

3

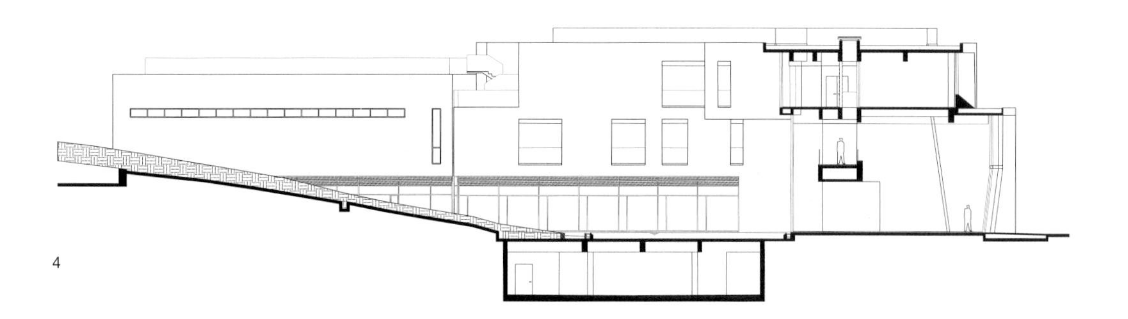

4

1/3 庭院以草坡及浅水景观池为两大主要元素，设计简洁，氛围静谧
With turf slope and shallow landscape pool, the courtyard enjoys a tranquil and neat environment

2 庭院设计概念草图
Concept sketch of courtyard

4 剖面图
Section

5-6 室外平台进深极大，压低了观赏视线，远方的山成了画面主角
The large depth of the semi-outdoor platform lowers the observation sightline and makes the mountain in a distance the key feature in the view

7 画廊式景框设计，把周边如画般的自然景色引入室内
Gallery type landscape frame introducing the beautiful landscape into the door

8 建设中的气象站
Meteorological station under construction

5

6

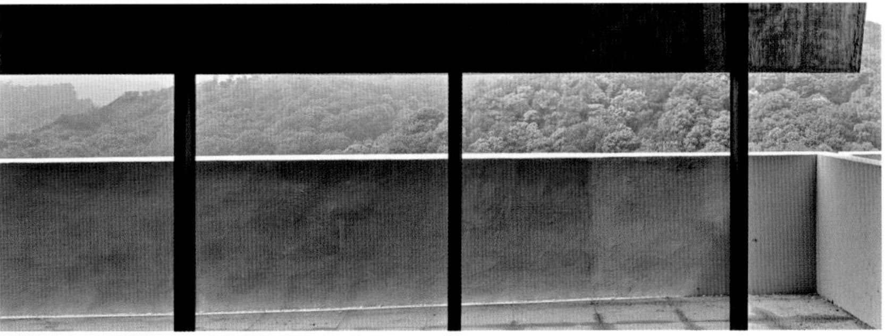
7

8

东莞勤上LED照明研发设计中心
Kingsun Led Lighting R&D Center,Dongguan

2010—　东莞 松山湖 / Songshan Lake Dongguan

2

LED照明研发设计中心是上市公司东莞勤上集团位于松山湖高新产业园的办公总部，是一组集办公、技术研发试验、展示展览等功能的综合建筑。无论在办公环境，还是建筑形象方面，本项目都体现了东莞勤上集团积极进取、敢为人先的企业文化。

项目建设将分为两个阶段，第一阶段的建设包括集团办公、实验室、展示厅等，第二阶段建设对外出租的办公楼及其配套设施。项目总图布局很好的平衡了功能要求、建设分期以及当地严格的城市规划管理条例等因素，合理规划建筑功能布局及首期建设的体量，利用架空层及内部环岛有效地组织对外来访、对内办公、后勤等流线。

本项目的一个极为重要的角色是集团展示其LED产品及研发成果的平台。在建筑内部设有一个跨越首层至五层的展示厅。展厅参观流线连贯，结合通高空间、天窗、缓坡道等，营造了独特的展览空间体验。另外，建筑造型将融合LED灯光设计，向城市展示勤上集团的企业形象，无论在白天，还是在夜晚，LED照明研发设计中心都能够成为耀眼的企业名片。

Situated in Songshan Lake Industrial Park, Dongguan LED Lighting R&D Center features a complex of buildings that houses office, R&D laboratory, showroom and other program spaces. The Center is also the place where the office headquarters of Kingsun Group, a listed company, is located. The Project fully showcases the aggressive and pioneering spirit of the company.

The construction of this Project consists of two phases. The first phase covers the construction of the headquarters office, laboratory and showroom, etc. The second phase includes the construction of office buildings for lease and their supporting facilities. The master plan has achieved a well-balanced layout in terms of functional requirements, construction phases and compliance with local urban planning regulations.

And more importantly, this Project will provide a platform for the company to display its LED products and R&D results. A five-story exhibition hall that runs from the first floor to the fifth floor will be arranged in the building. The LED lighting design will be integrated into the architectural form, thus making the LED Lighting R&D Center become a dazzling business card of the company all hours of the day and night.

项目地点：东莞 松山湖
设计时间：2010—2011年
建设时间：2011年至今
用地面积：2.3万m^2
建筑面积：约8万m^2
建筑层数：18层
建筑高度：75m
曾获奖项：2012东莞市优秀建筑工程设计方案一等奖

Location: Songshanhu, Dongguan
Design:2010-2011
Construction: 2011 to date
Site:23,000m^2
GFA: about 80,000m^2
Number of floors: 18
Building height: 75m
Awards: The First Prize for Excellent Architecture Design Award of Dongguan(2012)

3

1　总平面图
Site plan

2　LED照明研发中心东立面：东立面是研发中心重要的形象展示面。位于展厅下方的架空层是研发中心的贵宾入口，与其连接的内庭是整个项目解决复杂的流线关系的关键
East facade of LED Lighting R&D Center is an important display of the image of the center. The open-floor under the exhibition hall is VIP entrance of R&D Center. The inner court connecting such open-floor is the key to solve the complicate circulation relations of the entire project

3　LED照明研发中心主要沿街立面：建筑立面主要采用了大面的玻璃幕墙及竖向百叶，并采用不同色块的玻璃及铝板相互错落的处理手法创造丰富的立面肌理。道路交界处的草坡向城市开放，创造了积极的城市空间
Large glazing and vertical lamellas are employed in the building façade, where glass in different colors is staggered with aluminum panels to form diversified façade texture. The turf slope at the road crossing is made open to the city to create dynamic urban spaces

4　剖面图
Section

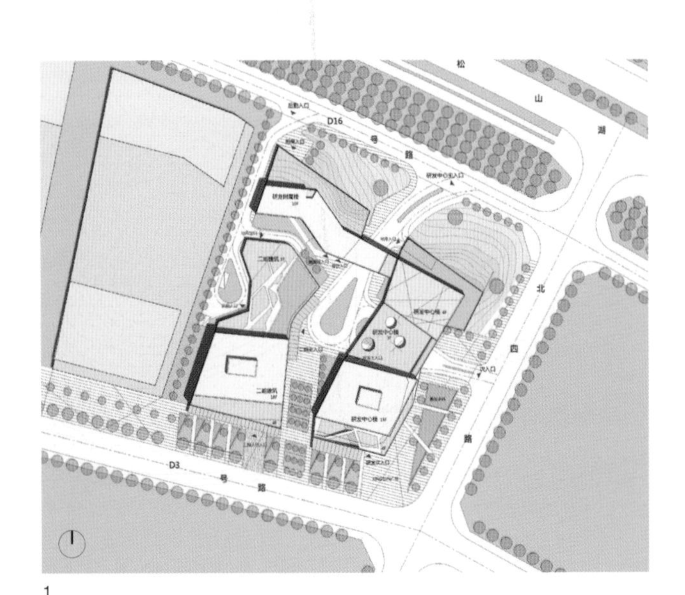

1

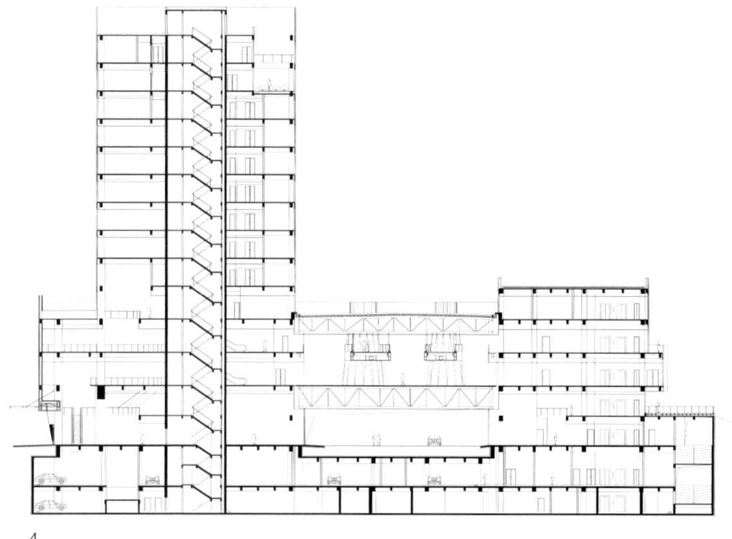

4

佛山欧浦国际商业中心
Europol International Commercial Center, Foshan

2013－　佛山 禅城 / Foshan Chancheng

欧浦国际商业中心地处佛山新城CBD东南角，视野开阔，景观优越，可达性及标志性俱佳。项目有两栋塔式高层办公楼，底部通过商业裙楼相互连接。在整体布局上，形成一个高低错落，遥相呼应的城市空间。在空间组合的基础上，合理地将各方向的人流组织起来，在整个区域大环境以及地块小环境中能合理地将地块功能充分发挥出来。

从城市天际线来看，为城市带来高低错落的效果，两座塔楼的东高西低，为达到视觉上的差异性，所以在设计上东西座高差超过30m。从裙楼上看两座分布并不对称，主入口幕墙通过刻意的偏移达到均衡的视觉效果，而不对称的外形反而更容易令人印象深刻。

双塔建筑外观表现为钻石形雕塑体块，通过精准工业手段在视觉上与简洁的艺术线条完美结合，体现项目业主的品质追求。顺滑流畅的曲线与强而有力的直线恰到好处地结合，整体形成协调的比例，建成的双塔在佛山新城区具有一定的标志性，与欧浦总部产生很好的呼应效果。

根据这种先天优势，在立面设计上尽可能利用体型的每一处表面：入口大堂采用通透的玻璃幕墙设计，增强外部的标识感和内部的通透感；办公室围绕塔楼核心筒布置，保证每个办公室有最大限度的视野面。

为达到佛山新城统一的规划要求，项目西座北地块设置大面积的中央景观绿化，并结合气候条件及周边环境，利用建筑物周边设置零星绿地和集中绿地，利用裙楼和塔楼屋顶设置屋面绿化，从而形成立体的，多层次的，点、线、面相结合的绿地系统，提供舒适、健康的办公环境，体现生态思想和节能理念。

Sitting on the southeast corner of CBD Foshan New City, Europol International Commercial Center embraces an extensive vision, excellent landscape, easy accessibility and recognition as a landmark. The project has two towers of office building, interconnected through business podium at the foundation. From the perspective of the master layout, it creates a layered and concerted urban space.

It provides a layered and coordinated effect to the city's skyline, as the east tower is higher than the west one. To create the visual difference, the height gap between the east tower and the west tower is over 30m. The curtain wall at the main entrance is deliberately shifted to achieve the balanced visual effect.

The outlook of the two towers is of diamond-style sculptor blocks. Through precise industrial approach, a perfect combination between visual effect and line art, revealing the pursuit of the owner. The two towers after completion will serve as a landmark in Foshan New City, echoing with the headquarters of Europol.

Each and every surface of the shape is to be used to its fullest in the façade design: transparent glass curtain wall is used at the entrance lobby to enhance the external identity and lighting within; offices of the tower are arranged centering the core tube to ensure each office's maximum view surface.

A huge central landscape planting is built in the northern plot of the west tower. A green space system will be created using the scattered green space and centralized green space around the buildings, roof green space out of the podium and towers to provide an office environment that is cozy and healthy, highlighting the eco thought and energy efficiency ideas.

项目地点：佛山市佛山新城
设计时间：2013年至今
建设时间：2014年至今
建筑面积：西座11.6万m^2+东座10.6万m^2
建筑层数：地上45层，地下4层
建筑高度：220.7m
合作单位：美国杰奥斯商业建筑设计

Location: Foshan New City, Foshan
Design:2013 to date
Construction: 2014 to date
GFA: 116,000m^2 for west tower + 106,000m^2 for east tower
Number of floors: 45 above-grade floors and 4 below-grade floors
Building height: 220.7m
Partner: GLC (USA)

1　总平面图
Site Plan

2　鸟瞰图
Bird's eye view

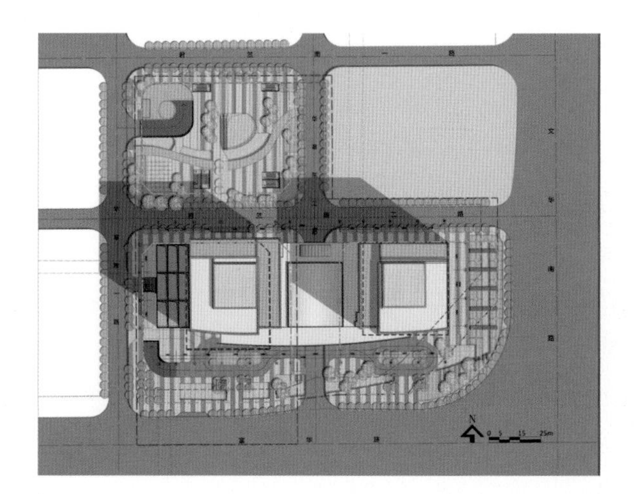

1

2

中山小榄金融大厦
Zhongshan Xiaolan Financial Plaza

2010— 中山 小榄 / Xiaolan Zhongshan

2

中山小榄金融大厦作为小榄镇的“水都发展核”，以融合“职、住、游的水都小榄镇”的中核分区发展为目标，对小榄镇乃至周边区域具有外在和内在的地标作用。该项目由一栋19层高的办公楼和一栋3层高的商业楼以及地下空间联合组成。

基于地区标杆性的定位，本项目采用了包括综合外遮阳系统、地下室光导照明、地面无机动车环境、双层立面，BIM手段，定位绿色二星标准等先进的设计标准和手段，突显其先进、环保、高效的特点。

办公楼有高大的体量，从人性关怀角度进行设计，并没有采用常见的外玻璃幕墙设计，而是采用有遮阳节能功能的外遮阳系统，一方面可以达到隔热目的（对内部使用人员的关怀），另一方面也直接减少光污染（对建筑外人流和其他建筑物内人员的关怀）。办公楼内方正实用，体现了对用地资源的节约。

商业楼只有3层，且处于内广场，因此采用了与办公楼截然不同的立面处理概念，采用外玻璃内窗中间为装饰铝条的双层立面手法，凸显商业展示性的特点，符合其为广场服务的本质功能。同时双层立面达到一定的隔热功能，为节能作出贡献。立面设计时考虑了与办公楼的呼应关系，采用了相近的比例分隔。虽然是截然不同的材料和方式，但两栋楼的立面设计非常协调、统一。

由于需要实现地面无机动车的设计，因此在本项目的地下室引入了VIP门厅的概念，将一般办公楼首层门前的VIP上落区设计在地下一层。由于有独立的分区设计，因此达到更好的VIP尊贵效果。同时也为节约用地作出了贡献。

Zhongshan Xiaolan Financial Plaza, as the core of the water town development initiative, will become a landmark of Xiaolan Town and beyond. Aiming to integrate the functions of “workplace, residence and tourism”, the project consists of a 19-floor office building, a 3-floor business building and underground space.

The project adopts leading design standard and methods including comprehensive external sunshade system, light-guided underground lighting, vehicle-free ground environment, double-layer façade, BIM approach, two-star green standard, highlighting its advancing, green and efficient characteristics.

Of a huge massing, the office building is designed to be attentive to human care. Rather than the common glass curtain wall, it use sun-screening and energy-efficient external sunshade system. It's a foursquare shape inside, reflecting its land utilization.

The business building has only 3 floors, and is inside a plaza. So a quite different façade treatment concept is adopted. The double-skinned façade has a certain thermal insulation, contributing to the energy efficient. Its correlation with the office building is taken into account during the façade design, and a proportionate compartmentalization is used, which makes people believe that the two are of quite different materials and forms, but inside the plaza they are well coordinated.

To ensure there is no vehicular traffic on the ground level, the VIP pick-up/drop-off foyer is placed on B1 instead of the typical ground level, which emphasizes the exclusive VIP service and contribute to more efficient land use.

项目地点：中山 小榄
设计时间：2010—2011年
建设时间：2011年至今
建筑面积：87420m²米
建筑层数：地上19层，地下2层
建筑高度：85m
合作单位：株式会社日总建

Location: Xiaolan, Zhongshan
Design:2010-2011
Construction: 2011 to date
GFA: 87420m²
Number of floors: 19 above-grade floors and 2 below-grade floors
Building height: 85m
Partner: NISSOKEN Architects/Engineers

1 东/西立面遮阳百叶角度分析图
Angle analysis of east/west façade sunshade louvers

2 项目地理位置优越，是当地的地标建筑之一
The Project is prominently located and planned as a landmark in the area.

3 从外部看办公楼，建筑物犹如穿了一件薄薄的轻纱一般。适应岭南气候，建筑南北立面与东西立面采用不同密度及方向的遮阳百叶
As if wearing a thin veil, the office building well adapts to the climate of Lingnan region, and is provided with sunshade lamellas of different densities and at various directions in the north/south and east/west facades respectively.

3

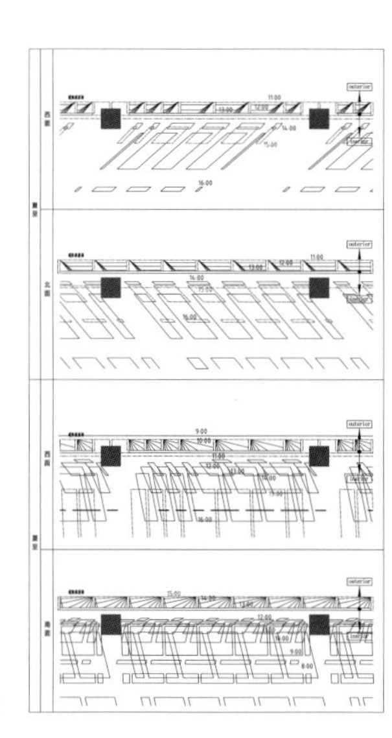

1

4

5

4 简洁的大堂空间，在迎客墙上用多种颜色陶棍组合就能形成印象深刻的空间效果
Concise lobby space. Ceramic rods of varied colors on the welcome wall create an impressive spatial effect

5 办公楼首层设计的休息厅，为紧张的金融行为创造舒缓的气氛，室外绿树与河景使人心旷神怡
The lounge on F1 of the office building, together with the exterior lush trees and river view, fosters a relaxing atmosphere for the intense financial workers

保利地产 · 珠海横琴发展大厦
Poly · Zhuhai Hengqin Development Building

2011— 珠海 横琴 / Hengqin Zhuhai

珠海市横琴岛地处珠江口西岸，与澳门隔河相望，是带动珠三角、服务港澳、率先发展的粤港澳紧密合作示范区。横琴发展大厦是横琴新区的启动项目之一，依托横琴岛“山水相融”的规划布局，以“打造横琴特有的城市与建筑风貌”为主要目标，围绕着三个最基本的要素展开设计：洁白光亮的建筑、遮阳的表皮与露台、风的竖井与架空层。

发展大厦塔楼方正，建筑高度达100m。位于二层的架空层及塔楼中央的“风的竖井”构成了建筑造型的最大特色，是实现其自然采光、通风的关键因素。架空层以下的建筑利用绿化屋顶做绿丘，延伸到南侧广场形成具有特色的大地景观。架空层以上的塔楼部分设置白色水平遮阳百叶，形成匀质界面。水平百叶后与错位排列的的开放式露台创造出轻盈灵动的建筑形象。塔楼内部功能布局集中，分区明确，流线明晰。核心筒四角均衡布置，使用者从各区域都容易到达目的地。平面空间可分可合，有利于适应不同类型的办公空间灵活使用需求。周边绿化露台与室内办公空间、公共空间形成不同的组合，提供了有趣味、舒适的办公空间。

发展大厦的设计尊重岭南气候特征，注重舒适、节能，体现其智能化办公建筑的理念。架空层与风的竖井相结合，风沿绿丘而上，促进塔楼自然换气。水平遮阳百叶表皮与露台相结合，能最大限度的隔绝夏日骄阳，降低建筑能耗。此外，还使用雨水收集系统、太阳能光电板和光导照明等多项绿色建筑节能措施。发展大厦将成为先进的节能建筑物，达到全生命周期节能的绿色建筑目标，实现其可持续性。

Hengqin Island of Zhuhai, located to the west bank of the Pearl River Estuary and just a river away from Macao, is planned as a pilot area for close cooperation among Guangdong, Hong Kong and Macao. As the kick-off project of Hengqin New Area, Hengqin Development Building aims to “create unique urban and architectural presence of Hengqin”. The design centers on three basic elements, i.e. “white and bright building, shaded skin and terrace air shaft and open-up floor”.

The square-shaped 100m Development Building will be occupied by the administration authorities of Hengqin New Area. The design aims to create a moderate, amicable, efficient, green and energy efficient building. The building is characterized by open-up floor, large cantilever and “air shaft” of central tower. The white horizontal lamellas clad on the tower above the open-up floor and staggered open terraces create light and flexible building appearance. The flexibly combined and divided planar space accommodates the demands of different offices, while the different combinations of greened terraces, interior offices and public spaces realize the interesting and comfortable office spaces.

Paying respect to the climate features of Lingnan Area, the design demonstrates the concept of intelligent office with consideration to comfort level and energy efficiency. As a cutting-edge energy efficient building, the Development Building can attain the green building target throughout the life cycle with ensured sustainability.

项目地点：珠海 横琴新区
设计时间：2011—2014年
建设时间：2011年至今
建筑面积：约23万m²
建筑层数：17层
建筑高度：100m
合作单位：日本佐藤综合计画
曾获奖项：2011年国际竞赛第一名
2014年广东省注册建筑师协会第七次优秀建筑佳作奖

Location: Hengqin District, Zhuhai, Guangdong Province
Design: 2011-2014
Construction: 2011 to date
GFA: about 230,000m²
Number of floors: 17
Building height: 100m
Partner: AXS
Awards:
Winning proposal of international competition(2011)
The 7th Excellent Architecture Creation Award by Guangdong Chapter of Association of Chinese Registered Architects (2014)

1 总平面图
Site Plan

2 景观绿化广场与低层绿化屋顶相连形成绿丘。绿丘之上悬浮着崭新的办公设施。开阔的开放性空间为公众活动提供优质的场所，大阶梯、二层天梯形成立体的多层次交通体系
The landscaped square and the green roof of the lower-rise are connected to form green mounds, with brand-new office facilities floating above. The open space offers quality venue for public activities, while the grand steps and sky stairs on F2 establish a multi-level traffic system

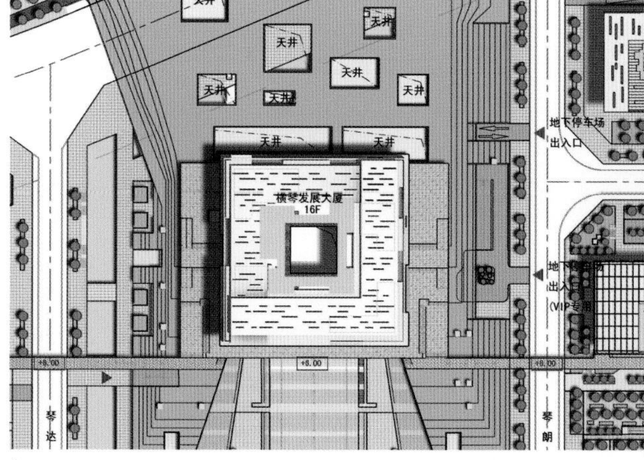

1

2

1

2

3

1 光塔式办公楼。夜晚，室内光线从遮阳百叶中透射出来，展现独具特色的反光效果。与广场、绿丘、水体的灯光相辉映，形成一番美景
Office buildings appear like light towers at night when the interior light penetrates from the sunshade lamellas, bringing about the unique reflective effect. Coupled with the light in the square, green hill and water body, a picturesque scene is then depicted

2 单体鸟瞰图
Bird's eye view of singular building

3 立面设计强调有横琴风格建筑特色，采用白色基调、白色的横向百叶、错位展开的露台、四个立面不分主次，形成轻盈活泼的建筑立面形象
The façade design highlights the architectural features of Hengqin style. The dominant white color, white horizontal lamellas, staggered terraces, and four equally important facades shape up an airily and lively façade image

4 塔楼中央竖向天井——"风之竖井"与二层开放式架空层，共同构筑风的通道，为建筑内部带来自然的风和盎然的绿意
Central shaft – "wind shaft" and the two-storey open-up floor jointly form a wind channel that brings natural wind and abundant green into the building

4

保利地产·珠海横琴保利国际广场
Poly · Zhuhai Hengqin Poly International Plaza

2013— 珠海 横琴 / Hengqin Zhuhai

本项目是由珠海横琴保利利和投资有限公司开发建设的大型办公楼和商业项目。设计重视使用上的便捷性和合理性，通过以人为本、绿色生态、形象优美独特、与山水格局相协调，形成建筑风格独特、色彩分明、功能完善、景观优美、品味高雅的城市中心区，打造"横琴风格、横琴色彩"的城市风貌。

建筑规划布局配合城市发展，形成良好的配套服务和功能互补，并充分利用新区的条件，规划组织为优质的办公和商业活动场所。本项目分设南北两座办公塔楼，商业裙楼居中布置并沿西侧道路南北展开，两座办公楼均是高层塔楼，从南向北依次升高，形成有序的天际线。南北两座办公楼均采用塔式设计，建筑平面布局功能明确、高效、协调、经济合理。建筑物南北间距约100m，并且紧临市政道路，视野开阔、交通方便，可以获得较好的景观、视线、采光和通风资源。

建筑位于横琴新区的中心地段，为横琴城市形象特色展示的重要区域，立面设计以鲜明的白色构筑基调色彩，以映衬玻璃的青辉和木质的本色，采用光感素材的玻璃质感映照周边美丽的自然环境，创造出柔和模糊的界面，使建筑与环境浑然融为一体。外墙分为"内表皮"和"外表皮"设计，内表皮采用玻璃幕墙，建筑内部人的活动可以积极地展现出来，增强建筑的活力和生命表现力，大片的玻璃可以为室内带来广阔的视野，并将充足的阳光与外部景观引入室内，提高室内环境质量。"外表皮"采用横向"钻石"形百叶形成遮阳系统控制阳光的照射，能有效防止热量的侵入，具有高节能效果；同时百叶的穿插变化，形成非均质的立面，创造出一个具有识别性的"横琴风格"。

As a mixed-use development for both office and commercial uses, Zhuhai Hengqin Poly International Plaza emphasizes the convenient and rational uses, meanwhile, through the human-oriented concept, green and ecological environment as well as elegant form, strives to shape up a fully-functional, attractive and trendy urban center with unique Hengqin style and color.

The Project is composed of a south and a north office tower, with a commercial podium in the middle stretching toward both towers along the western road. The building heights of the two said high-rise office towers ascend gradually from the south to the north to create an orderly skyline.

The façade design employs a vibrant white base color. Shimmering glass is used to reflect the beautiful surroundings and create soft interfaces that blur the boundary between architecture and the environment. The external wall is designed with both inner and outer skins. The former uses glass façade to vividly show the activity of people inside the building, ensure wide view into the outside and improve the indoor environment quality by introducing sufficient sunlight and exterior landscape into the building. The latter adopts a sunshade system comprising horizontal diamond-shaped lamellas to control sunlight and prevent heat from entering the building, thus reducing energy consumption. The interspersed and changeful lamellas form an inhomogeneous façade and eventually a highly recognizable Hengqin Style.

项目地点：广东省珠海市横琴
设计时间：2013—2014年
建设时间：2014年至今
用地面积：18276m²
建筑面积：10.24万m²
建筑层数：地上3-27层，地下2层
建筑高度：99.9m

Location: Hengqin, Zhuhai City, Guangdong
Design: 2013-2014
Construction: 2014 to date
Site: 18,276m²
GFA: 102,400m²
Number of floors: Above-grade: 3-27; below-grade: 2
Building height: 99.9m

1 总平面图 Site Plan
2 低点透视 Low view-angle perspective
3 裙楼透视 Podium Perspective
4 鸟瞰图 Bird's eye view
5 室外广场 Outdoor Square
6 大堂入口 Entrance Lobby

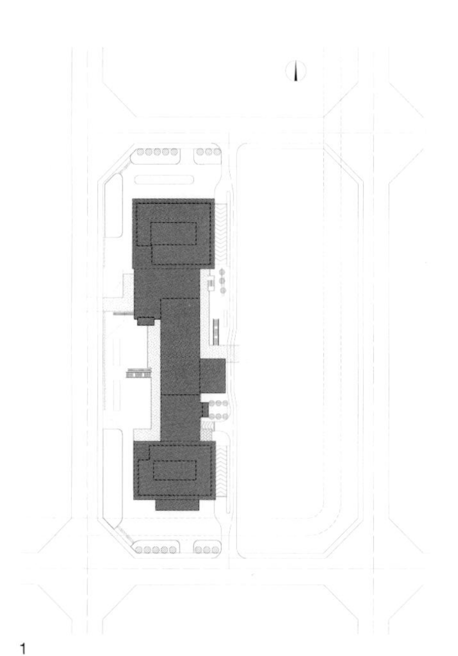
1

2

3

4

5

6

珠海横琴地方税务局新办公楼
New Office Building for Zhuhai Municipal Local Taxation Bureau

2012　珠海 横琴区 / Hengqin Zhuhai

本项目与横琴发展大厦相邻，属于同一城市设计范围。该区域的城市设计导则对于本项目的建筑高度、形态及规模都有指导性的意见。在满足城市设计要求的同时充分考虑具体办公楼功能需求，并创造现代行政办公楼新形象成为本项目设计的重点与难点。

城市设计中提出了塑造横琴风格的建筑元素：洁白光亮的建筑，遮阳的表皮与露台、自然通风和采光的竖井。这些元素均被应用于本办公楼项目中。开敞的绿化阳台，通高的中庭，丰富了其简单的方形体量，活跃了建筑的内外空间，提供了舒适办公空间。

白色的横向通长百叶作为遮阳的表皮，对于横琴有着琴弦合奏的寓意。办公楼的建筑立面处理采用了条状百叶，既呼应建筑的琴弦元素，为合奏唱响了伴奏，又使本建筑更有自己的个性。大楼未采用常规行政办公楼的对称形体，而是利用流动的条状百叶勾勒办公楼活泼的立面形象，同时也让立面上丰富的退台空间不显凌乱。这一活泼开放形象代表着行政办公楼的风格从严肃大方转向开放、亲民的趋势。另外每层楼外百叶的高度的设置也充分考虑到视线不遮挡的原则，让这一立面元素美观更实用，随着室外光线的变化，办公楼在早上和晚上都有丰富的建筑表情。办公楼东侧临近市民出入口处从三层开始设置了通高至顶层的中庭，改善普通办公楼通风采光不足的缺陷。中庭内部也不同于传统设计营造光洁的中庭内壁的手法，而是在中庭内壁上凸出许多“方盒子”办公空间，既增加了楼内办公空间的多样性，也让中庭空间更加有趣味。

The project is adjacent to Hengqin Development Building and falls within the same urban design unit with the latter. Satisfying relevant urban design requirements and functional needs of an office building and creating a modern administrative office building image pose a great challenge to the architectural design of the project.

Some architectural elements addressing the unique style of Hengqin have been defined in urban design, which are a white and shiny building, shading skins & terraces and ventilation shafts & elevated floors. They have been applied in the office building project. The simple square volume of the building ensures comfortable office space.

The white horizontal full-length fins that form the shading structure of the building resemble the strings of a Chinese zither ("Hengqin" in Chinese literally means zither). Stripy fins on the façade not only remind people of the zither strings but also form a unique character of the building. The flowing stripy fins create a lively and vibrant façade image and put the multiple recessed spaces on the façade in order. The building shows different appearances in daytime and at evening as outdoor daylight changes. The box-like office spaces that are projecting from the interior walls of the atrium not only increase the diversity the office spaces in the building but also make the atrium more intriguing.

项目地点：广东省珠海市横琴区
设计时间：2012年
用地面积：8600m²
建筑面积：12000m²
建筑层数：地上12 层，地下2 层
建筑高度：60m

Location: Hengqin District, Zhuhai, Guangdong Province
Design: 2012
Site:8,600m²
GFA: 12,000m²
Number of floors: 12 above-grade floors and 2 below-grade floors
Building height: 60m

1　中庭内壁上凸出的“方盒子”办公空间，既增加了楼内办公空间的多样性，也打破了原有中庭的简单、沉闷让中庭空间更加有趣味
The box-like office spaces protruding from the interior walls of the atrium not only increase the diversity the office spaces in the building but also make the atrium more intriguing

2　方案鸟瞰图，粗细变化的横向条状白色百叶呼应周边建筑，展现“横琴元素”的同时，也充分展示出自己的个性。不对称的活泼建筑形象更加凸显新区思维开放的政府职能部门亲民的办公服务态度
Bird's eye view. The white horizontal fins of varied sizes respond to the peripheral buildings, reflect the "Hengqin element", and fully showcase the unique character of the building. The asymmetric and lively architectural image further highlights the open mind of the government departments in the New Area and their amicable service attitude

1

2

珠海横琴新区横琴发展大厦建筑方案设计及周边地块城市设计
Architectural Schematic Design for Hengqin Development Building, Hengqin District, Zhuhai and Urban Design for Surrounding Plots

2011　珠海 横琴新区 / Hengqin Zhuhai

2009年8月，国务院正式批复了《横琴开发总体规划》，明确了横琴的战略方向——横琴成为"一国两制"下探索粤港澳合作新模式的示范区。城市设计的区域位于横琴新区核心地段，是以建设充满魅力的生态型观光都市为目标的示范区，将体现横琴的城市形象与文化品位，成为具有指导意义的优秀示范案例。

为了实现"横琴风格"这一目标，城市设计做如下考虑：1. 基于横琴新区"山脉田园、水脉都市"的城市发展理念，以叶脉状城市的概念展开规划设计。将城市市政设施、水路和道路像叶脉主干与枝干一样形成城市骨架和整体网络。水景、绿地与铺面做成交错变化的景观，渗透于城市之间。它是具有统一感的景观系统，并连接各建筑物，构筑起东西向穿插的城市网络体系。2. 将小横琴山与建筑用地区域内的景观相连为整体景观的一部分，发展大厦的用地即与山体绿地相连。山体绿地跨越主干道与河道延伸至建筑用地内，并向南侧港湾区续延。3. 设置富有绿意的都市公园。各都市公园通过水平方向的绿地连接形成绿色回廊。4. 各地块的建筑布局形成以都市公园为中心的中庭式格局，形成具有建筑围和感的城市公园。在各街区设置城市广场，形成各地块的入口，特色各有千秋。在各个街区重复这种格局从而形成横琴地区特有的都市风貌。5. 沿水路设置具有活力的亲水空间。主要道路旁设置空中廊道与面街建筑相连。6. 横琴发展大厦围绕洁白光亮的建筑，遮阳的表皮与露台，风的竖井与架空层这三个最基本的要素，展开一个高品位的建筑设计，打造具有横琴特色的建筑形态。区域内的其他建筑在统一格调中加以变奏。

横琴地区绿水葱郁，打造先进生态城市是本区域的重要目标。为实现环境优化组合，将点状独立的水域及城市公园中的绿地相互连接形成生态网络。整个城市是一个具备活力的生态系。将叶脉的生态轴做为生态网路的基干，水、绿、陆相结合形成独有的景观形态。由此衍生出富有魅力的生态景观，是只有横琴才能实现的具有示范性的城市案例。

In August 2009, the State Council officially approved the Master Plan for Development of Hengqin, specifying the strategy of developing Hengqin into a demonstration area for a new mode of cooperation between Guangdong, Hong Kong and Macao under the "One Country, Two Systems" policy. The urban design focuses on the core section of Hengqin New Area, intending to represent the urban presence and cultural taste of Hengqin, and provide an excellent case of guiding significance.

To achieve the target of Hengqin Style, urban design approaches below are considered. 1. The urban planning and design is based on the concept of a leaf veined city structure, where municipal facilities, waterways and roads form the city skeleton and the overall network. 2. Integrate Mt.Xiao Hengqin into the landscape of the construction site to make it part of the entire landscape. Spaces in front of Hengqin Development Building are in fact directly connected to the green space of the mountain. 3. Create a lush urban park. 4. The buildings of all plots center on the urban park to form a central-courtyard layout, creating a city park enclosed by buildings. 5. Create vibrant waterfront spaces along the waterway.6. The high quality architectural design of Hengqin Development Mansion is developed based on the three fundamental elements of a white and shiny building, shading skins & terraces and ventilation shafts & elevated floors. Other buildings within the area will follow the same style with minor modifications.

Hengqin boasts a luxuriant environment. Naturally, it is important to maintain its ecological features when building it into an urban district. In order to make the best of its natural environment, individual waters and green spaces in the urban park that scatter around like dots are interconnected to become an ecological network. The "leaf veins" are taken as the backbone of the ecological network in which waters, green colors and land jointly form the landscape that is one of its kind. Hengqin is an ideal area for a good demonstrative case for urban design.

项目地点：珠海市横琴新区
设计时间：2011年
设计范围：439000m²
用地面积：95851m²
建筑面积：331000m²
建筑层数：17-22层
建筑高度：100m
合作单位：日本佐藤综合计画
曾获奖项：2011年国际竞赛第一名
2013年度广东省注册建筑师协会优秀建筑佳作奖

Location: Hengqin District, Zhuhai, Guangdong Province
Construction: 2011
Urban design scope: 439,000m²
Architectural design site: 95,851 m²
GFA:331,000m²
Number of floors: 17-22
Building height: 100m
Partner: AXS
Award:
The First Place of national competition(2011)
Excellent Architecture Creation Award by Guangdong Chapter of Association of Chinese Registered Architects (2013)

1　整体设计概念
Master plan

2　叶脉状城市概念的规划设计，像蕉叶叶脉主干与支干一样形成城市骨架
The urban planning concept is based on the vein-like city structure

3　将山体的绿意引入用地区域
The greenery of the mountains is introduced into the site

4　整体鸟瞰图：各地块建筑布置，形成以城市公园为中心的中庭式格局，建筑与公园关系密切，形成整体感
Master bird's eye view: The buildings of all plots center on the urban park to form a central-courtyard layout, where buildings and park are closely related to and integrate with each other

5　天际线控制分析：设定临水近山的建筑高度较高，配合山体高度做天际线变化控制
Skyline control: allow for higher height for buildings near the water and mountain, and control the skyline variation in response to the mountain height

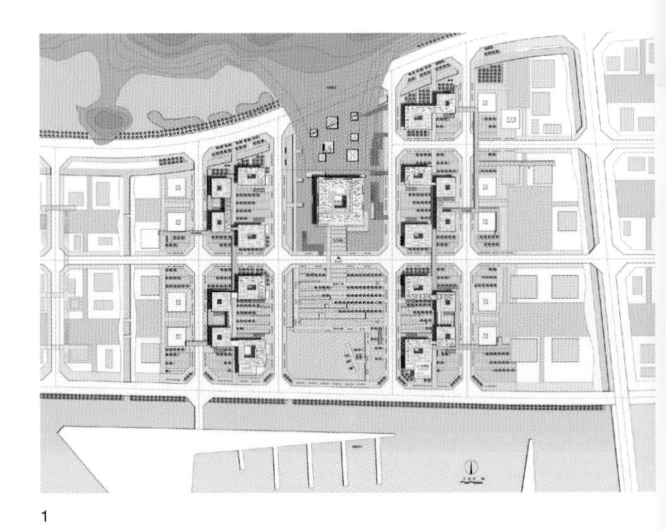

1

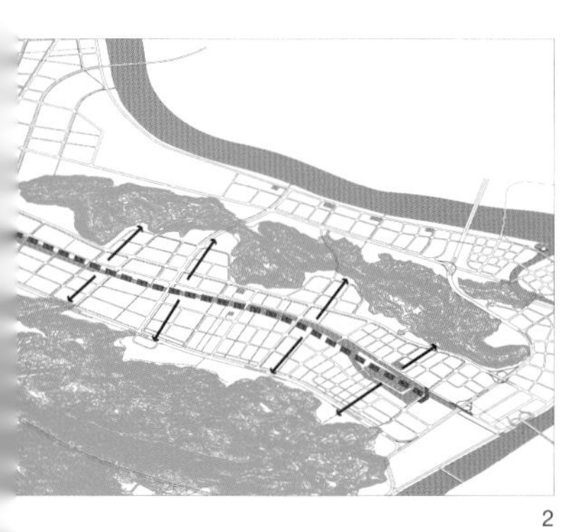

2

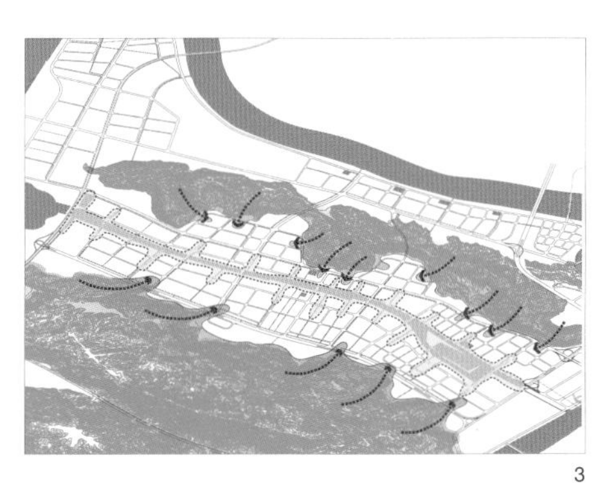

3

4

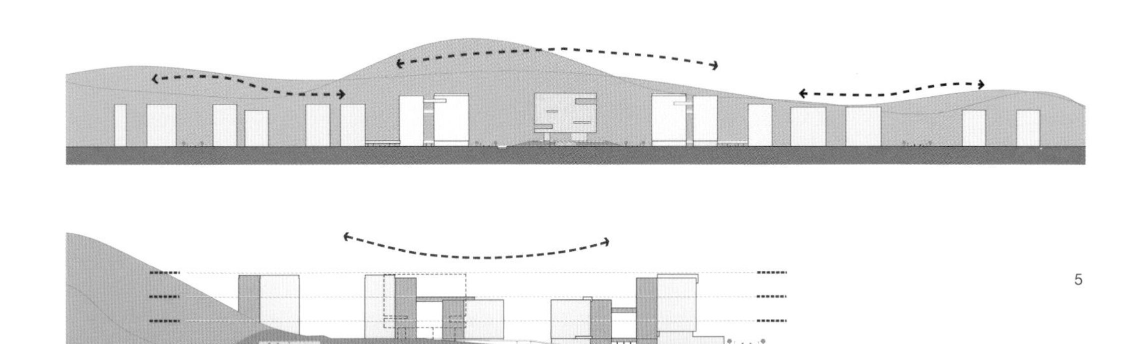

5

广州南沙新区蕉门河中心区城市设计暨建筑方案竞赛
Urban Design and Architectural Design Competition for Jiaomen River Central Area, Nansha, Guangzhou

2013　广州 南沙 / Nansha Guangzhou

南沙新区地处珠江出海口，是大珠江三角洲的地理几何中心。2012年9月，国务院正式批复《广州南沙新区发展规划》，作为南沙新区发展的纲领性文件，为本项目的开发建设提供有利契机。

设计上突破网格化布局的局限，使用地内的文化、科技、教育、商业等功能相互交织、融合互动，从而迸发活力，形成一个"链接的立体都市网络"。作为公共文化体育活动的核心区域，南沙滨水角文化体育中心区并非一览无遗的平面广场，而是一个层层铺开的城市"立体画卷"。为了实现城市区域的整体形象，避免由于多个公司参与公共建筑设计造成混乱的城市面貌，远期建筑单体的控制需要在如下框架下进行：一是公共建筑的主体部分设置在平台层标高上，形成飘浮感；二是需要强调楼板强有力的水平感，表现层级概念；三是文化建筑需要有独立的核心筒体量，形成贯入式核心体块；四是文化建筑入口层需要控制在二层，首层商业设施及地下一层车库需要与水体结合。

通过对能源利用的思考，我们考虑将大自然所赋予的"水·风·光·热"等能源零损耗地引入到人造的都市环境之中，并打造出具有场所特性的、自生自力的都市特点，这也是先进的生态都市所努力的目标。为此我们提出以"层级"的手法来实现环保型能源的统筹利用。即：地形层级（地形地势、山水地貌）、基础设施层级（能源、交通、公共设施基础网）、建筑功能层级（不同的建筑功能）。三个层级以两种秩序来构筑具体的都市形态：一是波纹状的构成：岭南丘陵地形、南沙有潮汐的河道都有波纹状曲线相类似的自然起伏，由此我们采用波纹状的网络曲线组织场地。二是放射形景观轴：以中央的开放空间为中心设置多种形态的放射形景观轴，规划设计"风·视线·活动"的通道。在整体控制下产生柔和的建筑，波纹母题既是建筑的围合也是行走的踏步，既是休息的平台空间也是人们观赏的风景，"水面""坡地""绿色"穿插在其间，形成绿水相融的风景。

In September 2012, the State Council officially approved the Development Plan for Nansha, Guangzhou which, as a guideline for development of Nansha, ushers in an unprecedented opportunity for this project.

The design ignores the constraints of the grid layout, and instead interweaves and interfuses the functions of culture, science and technology, education, business, etc. to work up the vitality and form a "articulated multi-layer urban network". Nansha Riverfront Cultural and Sports Center Zone is controlled within the following framework: firstly, place the main structure of a public building at platform level to maintain a "floating" feel; secondly, enhance the horizontality of the floor slabs to symbolize the hierarchy; thirdly, provide independent core for a cultural building; fourthly, place the entrance floor of a cultural building on 2F while integrating F1's commercial facilities and B1's garage with the river.

The "water, wind, light, heat" and other natural energy is zero-loss introduced to the built environment to establish the urban features that reflects the character of the place and highlights self-sustainability. Therefore, a "hierarchical approach" is proposed for utilization of the environmentally friendly energy. Such approach is composed of three layers, i.e. topographical layer, infrastructural layer, and architectural function layer. These three layers are arranged in two types of orders to shape the city. The first type is a wavy composition, while the second type is a radial landscape axis system. With such overall control efforts, the gently outlined buildings are created and the wave theme is reflected.

项目地点：广州 南沙
设计时间：2013年
整体范围：1658700m²
重点范围：559900m²
建筑面积：980038m²
建筑层数：2~32层
建筑高度：158.4m
合作单位：日本佐藤综合计画

Location: Nansha, Guangzhou
Design:2013
Overall urban design scope: 1,658,700m²
Key urban design scope: 559,900m²
GFA: 980,038m²
Number of floors: 2-32
Building height: 158.4m
Partner: AXS SAWTO Inc.

1

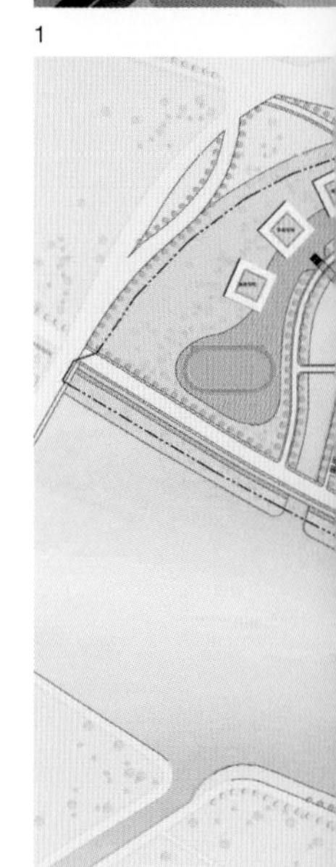

2

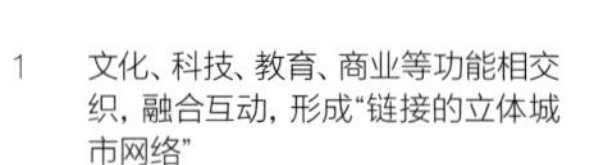

1　文化、科技、教育、商业等功能相交织，融合互动，形成"链接的立体城市网络"
The functions of culture, science and technology, education, retails, etc. are interwoven and integrated to form an "articulated multi-layer urban network"

2　总平面图
Site Plan

3　以大剧院为前景的滨水城市景观
Waterfront cityscape with the Opera House as the foreground

3

1

2

3

1　远期建筑单体的控制强调楼板强有力的水平感，表现层级概念
Control of the long-term singular building focuses on the powerful levelness of floor slab, to express the concept of layering

2　规划用地的中心，设计人群最易活动聚集的开放空间，以波纹状设计地形、基础设施、建筑功能，营造波纹状的基本网络
Center of the planned land is designed as an open space where people gather conveniently. Landform, infrastructural facilities and building functions are designed into wave shape to create a wavy elemental network

3　文化建筑面向蕉门河，与其后方的商业建筑围合出中央开放空间。与丘陵、河岸呼应的场地设计，以及与地下建筑相连的下沉广场，营造出层层铺开的，立体的城市空间
The cultural buildings facing Jiaomen River, together with the commercial building behind them enclose a central open space. Site design corresponding to the hills and river bank, and the sunken square connecting to the underground buildings forms a multi-level urban space

4 文化建筑入口层需要控制在二层，首层商业设施及地下一层车库需要与水体结合
The entrance of the cultural buildings is placed on F2, and the design of the F1 commercial facilities and B1 garage is considered in combination with water body

5 "水面""坡地""绿色"互相穿插，形成绿水相融的风景
Interspersed "waters", "slopes" and "green spaces" creates a landscape of green and water

4

5

岭南V谷-广州国际智能科技园项目
Lingnan Valley Guangzhou International Intelligent Science Park

2013－2014　广州 荔湾/ liwan Guangzhou

项目地点：广州市 荔湾区
设计时间：2013－2014年
建筑面积：54.17万m^2
建筑层数：地上33层，地下3层
建筑高度：最高150m

Location:Liwan District, Guangzhou
Design:2013-2014
GFA:541,700m^2
Number of floors:33 above-grade floors and 3 below-grade floors
Building height:150m

岭南V谷是广钢开发区片区地块作为广佛同城和白鹅潭经济圈的重要节点，广州钢铁集团计划依托广州市"三旧"改造政策，对该地块进行综合开发利用。目前地块已纳入广州市旧厂房改造规划范围。本项目是广钢构建高新技术产业体系加快推进战略转型的重要步骤，具有"先行先试"的示范效应。

岭南V谷是以科技为主题，以高新技术产业为指导，以商务办公为支撑，融合科研、商业、服务、休闲、娱乐于一体的科技主题RBD。园区着力培植智能装备、新材料和信息技术等国家战略型新兴产业、吸引科技型企业入园创业发展，建成具有国际影响力的科技创新、科技合作及科技服务的示范基地。作为科技园区，应兼顾效率、交流、形象、低碳等标志性形象，体现创新性的构思特点。构思将场地的折线形态特征进一步整合，形成贯穿始终的设计语言。

绵延江岸1.5km长起承转合的绿轴与庭院空间， 绿色、生态为主题的科技园区有更多滨江特色，折线形三维轮廓线层层错落，建筑设计引入"立体城市"（Multi-Ground City）的设计理念，探索在土地紧张，高强度开发前提下商业、研发、居住环境协调平衡的建筑模式。临江第一排以星级酒店为引领，后接裙房设置广钢风情情景商业街，与集中商业、休闲商业和沿街商业有机结合，充分利用沿江资源，完善园区配套，为园区添加特有吸引力和竞争力，提升园区价值品位。 建筑群呈现北高南低的整体趋势，其中西北角超高层写字楼为场地制高点。建筑群通过板点结合的模式，形成更宽的视线走廊，让第二、第三排的产业楼及办公楼能看到江景，并且建筑屋顶通过各种切角的处理变得更多变化，避免生硬的折线。重点运用"转"、"回"、"承"、"顺"的构图手法，在重点区域通过线性的收放，打造不同尺度、不同规模的活动空间，同时，在二维构图的基础上进行了三维空间的景观设计，在大尺度的线性尺度中设计小尺度的景观节点，通过这种无边界、无界定感的环境设计，给观者、游者持续的新奇感。

As a key node at Guangzhou-Foshan Area and Bai E Tan Economic Circle, Lingnan V-Valley represents the redevelopment and reuse initiative of Guangzhou Iron and Steel Enterprises Group under the city's "three-old redevelopment" policy. It is also an important step and a pilot project for the client to construct hi-tech industry system and facilitate the strategic restructuring.

As a science & technology themed RBD that integrates scientific research, commerce, service, recreation and entertainment, the project aims to incubate the country's newly rising strategic industries including the intelligent devices, new materials and information technology, attract the S&T enterprises and create a technological innovation, cooperation and service base that could stand out as example of international influence. The design concept for such an S&T park should incorporate the iconic images like efficiency, communication, low-carbon etc. The form of the site that resembles a set of broken lines is abstracted to become the design language used throughout the design process.

The 1.5-km-long waterfront green axis and courtyard spaces wind along the river, adding more waterfront features to this green/eco-themed park. With the staggered 3-D outlines of the broken lines, the multi-ground city concept is incorporated into the design to explore the building mode where commercial development, research facilities and living environment could be well balanced against the tight land use and high development intensity. The first-tier waterfront is dominated by the starred hotels. The podiums on the back house some featured shopping streets which well complement with shopping center, leisure commerce and street-front shops to perfect the supporting facilities, enhance the park's competitive edge and value. Building clusters witness a descending building height from north to south, with the office super-high rise in the northwest corner as the highest point. Building layout features the appropriate combination of bar-type and point-type buildings to allow for attractive river view to the industrial buildings and offices in the second/third tier waterfront. Activity spaces of varied sizes and scales are created through the various composition approaches and the linear treatments at key nodes. Meanwhile, 3D landscaping is provided on the basis of 2D composition and small scaled landscape nodes are created within the larger linear scale, which keeps bringing nice surprises to the visitors.

1

2

1　岭南V谷建筑沿珠江层层展开，错落有致，在夕阳下形成生动的立体图景
The buildings in Lingnan Valley extend elegantly and gradually along the Pearl River, depicting a vivid multi-level view in the setting sun

2　总平面图
Site Plan

3

4

5

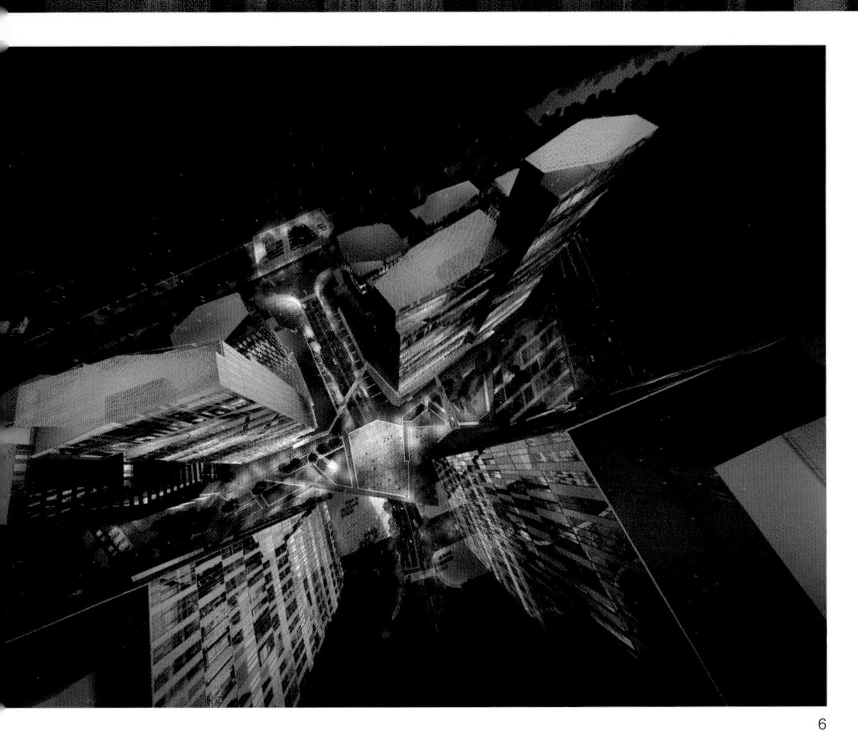

6

7

3 以场地固有折线特征为基本元素，控制建筑、景观、空间等形态
The inherent fold line feature of the site is taken as the basic elements to control the architecture, landscape and space patterns

4 滨江商业带与滨江绿地、邮轮码头形成整体休闲氛围
The riverfront commercial zone fosters an integral leisure atmosphere with the riverfront green space and cruise terminal

5 建筑群以轻松的姿态面向珠江口航道
The building cluster faces the Pearl River Estuary channel in a relaxed gesture

6 雨夜中的岭南V谷建筑群
The building cluster of Lingnan Valley at rainy night

7 以外廊为主的情境商业更容易形成热闹的休闲商业氛围
Context-based retails concentrated on the veranda make it easier to foster a busy yet leisure commercial atmosphere

广州海珠区龙潭村三旧改造项目概念方案
Conceptual Scheme for Longtan Village Redevelopment Project, Haizhu District, Guangzhou

2010-2011　广州 海珠 / Haizhu Guangzhou

在快速城市化带来的用地扩张背景下，根据上层次规划明确的发展方向和定位要求，龙潭村以居住和商业为主导功能，承接广州市居住功能分流，对中轴线使馆办公区的服务拓展，同时该规划平衡发展房地产业，积极发展商贸旅游业，积极利用海珠湖和万亩果园的景观资源，将旅游业与商业良性结合，相互促进。作为三旧改造在满足居住及配套功能的基础上，结合广州市的有关政策规划和市场要求，赋予部分用地新的商旅功能，充分体现其土地价值。

龙潭村改造的目标是改善居民居住环境，促进村集体和经济发展，提升地区整体品质。改造以控规为指导，合理布局村民、居民住宅和村集体物业用地；利用城中村土地收益增量支持改造，保障村民、居民和村集体的合法经济利益及发展权益，传承历史文化，延续文脉。

龙潭村未来功能发展主要方向为：优先满足村民、居民回迁的居住需求，改善人居环境，增强居住配套，同时以龙潭村河涌、海珠湖、万亩果园作为依托，沿河涌发展居住、旅游、商业服务功能，其中商业着重考虑酒店、公寓、写字楼、商场、零售百货超市等。

本次三旧改造范围总用地面积 72.12hm²，净用地面积（扣除规划确定的城市道路、高速公路及立交、市政设施、防护绿地用地外的全部用地） 48.48hm²；规划建筑面积总量约为224.46万m²，改造范围总体毛容积率为3.13，净容积率为 4.65。

从总体上看，改造后龙潭村建筑密度下降为32%。整体环境得到质的提升，旧村转变为配套服务齐全的现代城市社区。

It has been pointed out in upper level planning that Longtan Village will be mainly for residential and commercial uses. It will share some of the residential function of Guangzhou and serve the consulate area along the central axis of the city. Also, the planning will find a balance between property development and the development of commerce, trade and tourism. Redevelopment of the village will provide new business and tourism functions to the site to reflect the land value, provided that the residential and supporting uses are satisfied first.

Redevelopment of Longtan Village aims to improve the living conditions, promote its economic development and enhance the overall quality of the region. Guided by the regulatory detailed planning, a reasonable layout for land uses will be developed. Also, additional land revenue gained by the village will be used to support redevelopment and protect the rightful interests of the relevant parties.

In the future, the Longtan Village will give priority to the resettlement demands of villagers and residents, improve the living environment, provide more residential facilities and develop residential, tourism and commercial facilities along the canals, Haizhu Lake and Great Orchid. As for commercial facilities, the focus will be on hotels, apartment buildings, office buildings, shopping malls and supermarkets.

The redevelopment project covers a total site area of 72.12 hectares, including a net usable area of 48.48 hectares. The planned floor area is about 2.2446 million sqm with the gross FAR at 3.13 and the net FAR at 4.65.

With the building coverage of Longtan Village after redevelopment reduced to 32%, the overall environment has been improved and the old village transformed into a fully-functional modern urban community.

项目地点：广州市海珠区
设计时间：2010－2011年
用地面积：721200m²
建筑面积：2254600m²
建筑层数：最高65层

Location: Haizhu District, Guangzhou
Design:2010-2011
Site:721,200m²
GFA: 2,254,600m²
Number of floors: Max.65 floors

1　城市中轴线的又一个支点
Another pivot on the city's central spine

2　总平面
Site plan

3　与周边海珠湖公园和万亩果园遥相呼应
Longtan Village echoes with Haizhu Lake and Great Orchid in a distance

4　简洁明晰的造型，通透的幕墙，辉映万亩果园的湖面
The concise and clear-cut building form and the transparent façade are reflected in the lake of Great Orchid

5　居住区祠堂、古建结合村落的河涌蜿蜒布置，延续龙潭村历史文脉
The ancestral hall and ancient buildings are located along the meandering river, passing the historical and cultural heritage of the village on to future generations

1

2

3

4

5

广西城市规划建设展示馆

Guangxi Urban Planning and Construction Exhibition Hall

2009　广西 南宁 / Nanning Guangxi

广西城市规划建设展示馆的功能定位为广西城市规划和城市建设成就的展示平台、广西城市文化与城市历史的研究中心、广西城市规划和建设人才的学术交流中心。展示馆的建设地点位于五象森林公园内，北临邕江，周边环境优美。

考虑现有坡地地形，建筑布局依山就势，采用条状平行式布置，主立面沿城市景观轴线方向舒缓伸展，凸显其作为展览建筑的公共性。为呼应周边山地景观，建筑体量上下错动一层，与地形契合，大大地减少了填挖土方对环境的影响。不同标高的屋面如折板般连贯起来，整个建筑犹如从自然山丘上裸露出的一颗宝石。传统建筑的屋面，侧墙，天窗等无法承载浏览活动的建筑构件，被精心设计的多条坡道有机结合成一条连接室内室外以及不同标高的展厅浏览路线。无论是参观或者游玩，都能获得全新的观展体验和空间体验。

建筑体内部的交通流线围绕贯穿建筑体量的十字形中庭空间展开，在中庭空间中设置了多种垂直交通设施，包括电梯、自动扶梯和楼梯，给观众提供多种选择。同时还考虑了残障人士的交通方便，垂直电梯可到达每个展示楼层。沿中庭的长轴方向，设置了直跑楼梯，联系各个楼层的展厅；中庭两侧的展厅之间又有空中廊桥相互联系，形成连贯通畅的游览观展的循环路线。

Located in the picturesque Wuxiangling Forest Park, the project is planned as the province's platform for displaying the urban planning and construction achievements, a research center for urban culture and history, and an academic exchange center for the urban planning and construction professionals.

The buildings in stripe layout echo with the hilly terrain. The main façade extends along the city's landscape axis to highlight its nature as a public building. Meanwhile, the building volumes are staggered by one floors height to fit into the landform and largely reduce the impact of excavation and back filling on the environment. The whole building seems like a gem naturally exposed on the hills. The traditional building components like roof, sidewall and skylight that are not meant for visiting are integrated by multiple well-designed ramps into a visiting tour to connect interior/exterior spaces and exhibition halls at different levels. Visitors may either enter the exhibition halls from the square, or stroll along the slopes and among the roof-top exhibition areas at different elevations, enjoying brand new visiting and spatial experiences.

The circulation inside the building is organized around the cross-shaped atrium where various vertical transportation facilities are available. Along the long axis of the atrium, a straight run stair is provided to connect all the exhibition floors. The exhibition halls on either side of the atrium are connected with sky corridors to form a smooth and closed visiting route.

1

项目地点：广西 南宁 五象森林公园
设计时间：2009年
用地面积：6.66万m²
建筑面积：4.4万m²
建筑层数：地下1层，地上4层
建筑高度：25m
曾获奖项：2010广东省注册建筑师优秀建筑创作奖

Location: Wuxiangling Forest Park, Nanning, Guangxi Province
Design:2009
Site:66,600m²
GFA: 44,000 m²
Number of floors: 4 above-grade floors and 1 below-grade floor
Building height: 25m
Award:
Excellent Architecture Creation Award of Guangdong Chapter of Association of Chinese Registered Architects (2010)

3

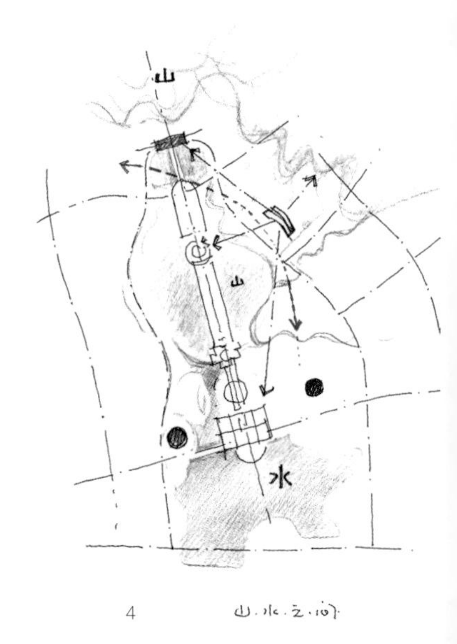

4　山·水·之·间

2

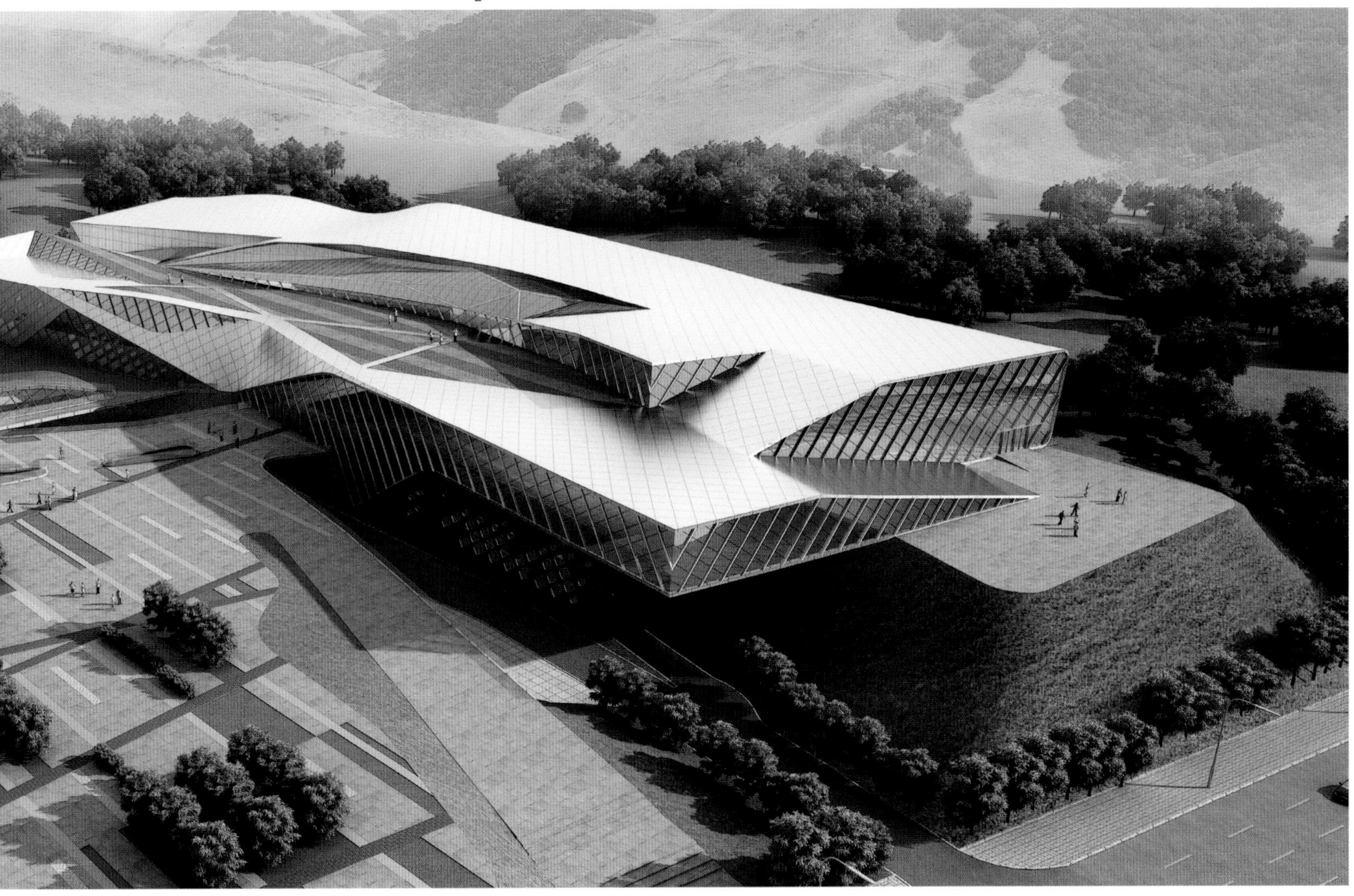

1 总平面图
Site Plan

2 夜景效果图
Night View Rendering

3 鸟瞰图：展厅或在山坡之下，或在屋面之上，或在建筑之中，观感丰富。大量折线和折面的光洁外壳让建筑犹如山丘上裸露出的一块璀璨的宝石
Bird's eye view: the exhibition halls below the hills, above the roof or in the buildings offer varied building images. The bright and clean shell comprising a great number of folded lines and surfaces makes the building a brilliant gem exposed on hills

4 景观分析图，建筑在总图上呈弧线蜿蜒，是为了让其两端的开窗能够远眺到景区的景观节点，欣赏展品的同时也能欣赏景区风景
Landscape analytical diagram: the winding buildings on the master plan allow visitors to enjoy the landscape nodes from the windows at both ends during their visits

5 设计手稿
Design sketch

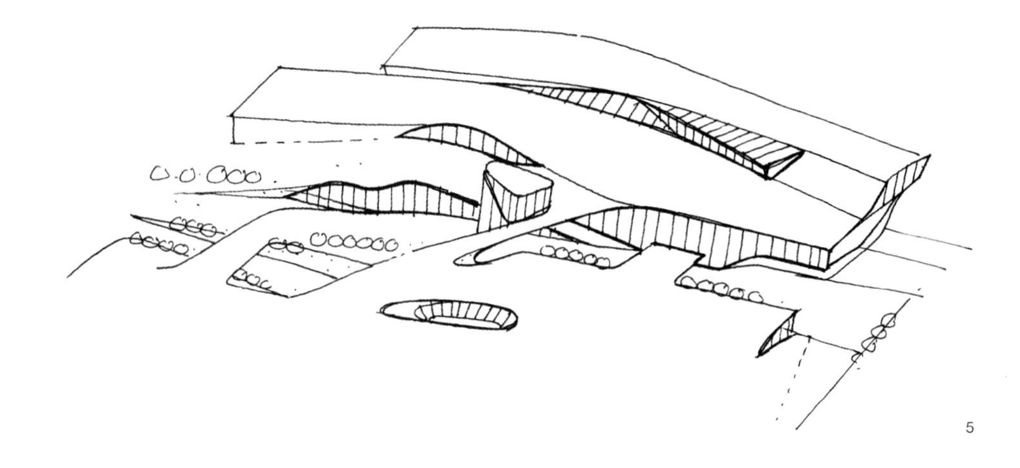

5

1

2

1　设计提案将一部分建筑体量嵌入山体之中，形成可上人屋面，从而达到了建筑与场地环境浑然一体的效果
The design sets part of building volume into the mountain massif to create an accessible roof, thus integrating the building naturally with the site environment

2　建筑面向城市展开，凸显其公共建筑的特性
The building stretches toward the city, highlighting its feature as a public building

3　设计手稿
Design sketch

4　内堂效果图
Inner Lobby Rendering

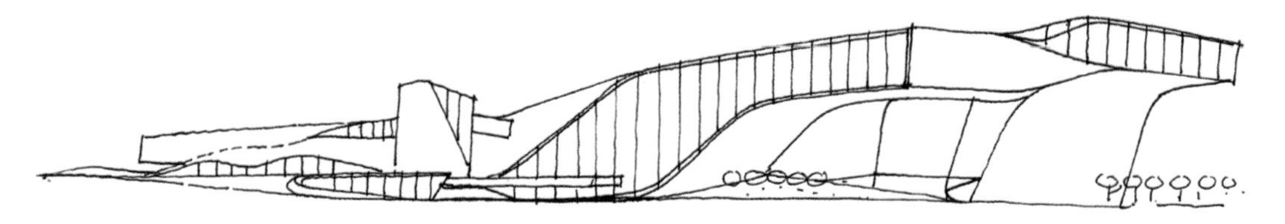

3

4

南方广播影视创意基地（一期）
Creative Base (Phase I), Southern Media Corporation

2012　广州 天河 / Tianhe Guangzhou

随着广东建设文化强省的号角吹响，南方广播影视创意基地成为广东省建设文化强省的十大精品项目之一。第九届全运会及第十六届亚运会的成功举办，也推动了以奥体中心为核心的城市东部新区的发展。在环境优美的奥体公园北侧，一个集办公、影视、动漫节目制播、外景拍摄、旅游购物、培训展览、商务酒店等功能于一体的"中国好莱坞"将拔地而起。它将成为广州的新地标，彰显新世纪广东广播电视行业的新面貌。

本项目的方案设计必须满足原有的规划设计的基本框架。设计方案通过整合不同单体的体量、造型，统一建筑设计手法加强建筑群的整体感。整体交通规划利用地形自然形成两个相差6m的台地区分组织场地内外各类流线。内部功能设计必须满足集团现在及未来影视节目制作的要求，各类演播厅竖向组合，集中布置，内部工作流线便捷明晰。同时，空间设计也要考虑到参观或参与节目录制的公众，在保证功能合理布置及面积节约的前提下，充分满足公共空间的舒适性。酒店及会议中心设计根据其用地及功能特征，整合一体化，引入可方便组合联系的大堂及交通流线，两者可分可合资源共享。

建筑造型采用现代简洁的表达方式，引入非线性设计，产生丰富变幻的视觉效果。位于前广场的剧院造型更加具有创意及动感，动感的建筑造型带来了同样流动的室内空间。多媒体展厅结合参观流线形成独特的造型效果，别致的流线设计使人们可以由地面通过坡道自然地走到屋顶花园，带来新颖的参观体验。建筑中引入屋顶花园，采用自然地貌的设计手法，花园自然融入形体变化中，成为建筑的一部分，随处可达，提升了建筑环境品质。办公楼立面使用可呼吸式双层幕墙，根据需要调节入射光线及引入新风，兼顾了办公区的景观效果和空间舒适度要求。同时在立面设计中运用彩釉玻璃的不同透明度的设计组合，形成丰富的肌理变化，降低遮阳系数，节能环保。

To promote the cultural undertaking in Guangdong, the Creative Base of Southern Media Corporation is developed as one of the ten key projects for this purpose. Just to the north of the Olympic Sports Center, the "Hollywood of China" is developed to integrate the office, production and broadcasting of film, TV and animations, location shooting, tourism, shopping, training, exhibition, business hotel and other functions. It will set a new landmark and promote the TV and broadcasting in Guangdong.

The project design, by integrating different volumes and shapes of individual buildings and unifying the architectural approach, enhances the sense of wholeness of the ensemble. The overall traffic plan centers on the two pieces of terraced lands at different heights to organize the internal and external circulation. The broadcasting studios are centralized and vertically stacked to ensure a clear-cut and convenient internal circulation. In spatial design, the visitors and participants to the programs are considered, while the public spaces are comfortable with rational layout and floor efficiency. The hotel and the conference center may share or separately use the various resources thanks to the flexible lobby and traffic circulation.

The modern and simple building form plays on the non-linear elements to generate constantly changing visual effect. For better environmental quality, roof garden is introduced to and naturally integrated with the changing building form. The office building employs breathable double-skin curtain walls, while the daylight and fresh air are introduced into the building as needed to realize desired views and comfort level in the office area. The combinations of fritted glass with varied transparency levels contribute to the diverse building fabrics and desired shading, energy efficiency and environmental protection effect.

项目地点：广州 天河
设计时间：2012年
用地面积：23.29万m^2
建筑面积：54.4万m^2
建筑层数：12层
建筑高度：50m
曾获奖项：2012年国内竞赛中标
2014广东省注册建筑师协会
第七次优秀建筑佳作奖

Location: Tianhe District, Guangzhou
Design:2012
Site:232,900m^2
GFA: 544,000m^2
Number of floors: 12
Building height: 50m
Awards:
The First Place of national competition
The 7th Excellent Architecture Creation Award of Guangdong Chapter of Association of Chinese Registered Architects (2014)

1

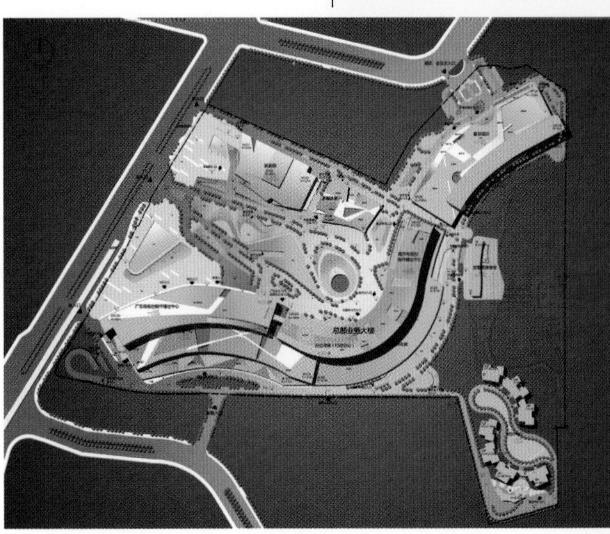
2

3

4

1　创意基地整体体量顺应地形，形成不同高差的两个台地，自然区分了内外广场
The Base follows the terrain to form two terraces at different height levels, thus naturally separate the internal and external squares

2　总平面图
Site Plan

3　建筑立面根据朝向采取了不同的处理手法，达到生态节能的目标
The building façade is treated differently to respond to the orientation and realize the ecological and energy saving goal

4　建筑造型采用现代简洁的表达方式，引入非线性设计，产生动感的视觉变化
Modern and simple architectural form and non-linear design, creating dynamic visional change

1

2

3

4

1 酒店大堂室内透视
Interior perspective of hotel lobby

2 歌剧院观众大厅室内效果图
Interior rendering of opera house auditorium

3 会议中心大厅室内效果图
Interior rendering of conference center hall

4 多媒体展厅结合参观流线形成独特的造型效果，别致的流线设计使人们可以由地面通过坡道自然地走到屋顶花园，带来新颖的参观体验
Multimedia exhibition hall is designed with a unique shape in combination with the visiting circulation. Unconventional circulation design allows people with easy access to the roof garden from the ground via the ramp, which provides original visiting experience

5 剖面图
Section

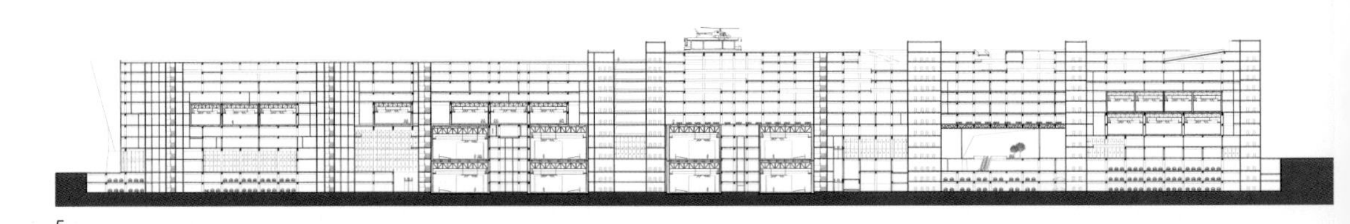

5

影剧院入口空间开出了若干中庭，增加不同公共区层间的交流，也引入了特意设计的天光。
Multiple atriums are provided at the entrance space of the theatre to enhance the inter-floor communication of various public areas and bring in the daylight.

创意基地源于一片绿地，建筑的置入与环境进行了很好的结合，将绿化引入空间各处，尊重自然。
The Creative Base is originated from a green space. The embedded buildings are well integrated with the environment, where green spaces are introduced into various spaces to show respect for the nature.

中新（广州）知识城规划展示厅
Exhibition Hall, Sino-Singapore Guangzhou Knowledge City

2010—2011　广州 萝岗区 / Luogang Guangzhou

萝岗是广州市的建设热点区域之一。中新知识城展示厅是萝岗招商引资、规划展示的重要平台，它包括了一系列功能，如规划展厅、会议接待中心及配套建筑、驿站等。建设地点位于萝岗区九龙大道以南，面向城市干道交叉口，用地西面有自然山体绿化，北面为自然土丘，周边环境优美。

项目采用合院式布局，展厅、会议中心围绕中央庭院周边布置。各部分功能空间既有联系又可独立运营，使用高效便捷。无论从室内还是户外，中央庭院是整个展示厅的重要景观。从城市主干道望去，建筑主立面横向展开，舒展大气。整体建筑造型简约，通过利用钢、木、玻璃、青砖这些材料的巧妙组合，体现建筑的独特性格。轻盈的钢结构玻璃雨棚、厚实的砖墙、通透的玻璃幕墙、柔和的实木百叶，既体现建筑的现代性，又富有岭南建筑的传统韵味。

展示厅采用造价适宜的节能方式，合理降低建设投资及建筑运营、维护费用。立面采用多样化的遮阳方式，从而减少建筑能耗；结合错落的庭院，场地中央设有浅水池，植物疏密有致，营造良好的景观，改善建筑微气候；屋面使用太阳能电池，为内部运营提供干净的能源。以上这些措施使建筑达到可持续发展的目标，也契合了萝岗区高技、环保的发展方向。

Prominently located to the south of Jiu Long Da Dao, Luogang District, Guangzhou, the Project faces the intersection of urban main roads. It enjoys nice views with the hills on the west and north. Planned with multiple functions like planning exhibition hall, convention and reception center, supporting building and post house, the project boast an important platform for Luogang to attract investors and show the district planning.

The central courtyard is framed by the exhibition hall and convention center. The functional spaces are connected yet allow for independent operation. The concise building shape and the ingenious combination of steel, wood, glazing and grey brick present the unique building identity. With light glazing canopy with steel structure, solid brick wall, transparent glass curtain wall and gentle wood lamella, this modern building also throws out the traditional flavor of Lingnan Architecture.

The cost-effective energy efficient measures duly cut the cost of construction, building operation and maintenance, while diversified façade shading devices reduce the building energy consumption. Those measures realize the sustainable development of the building and respond to the hi-tech and environment friendly development orientation of Luogang District.

项目地点：广州萝岗区九龙大道
建设时间：2010年（展厅一期）
　　　　　2011年（展厅二期）
用地面积：18900m²
建筑面积：7885m²
建筑层数：2层
建筑高度：12m

Location: Luogang District, Guangzhou
Construction: 2010 (Exhibition Hall – Phase I)
　　　　　　2011 (Exhibition Hall – Phase II)
Site:18,900m²
GFA: 7,885m²
Number of floors: 2
Building height: 12m

1　中央庭院是整个展示厅的重要景观。从城市主干道望去，建筑主立面横向展开，舒展大气
The central yard is the important landscape of the entire exhibition hall. Viewing from the urban trunk line, the main building façade stretches generously in a transverse manner

2　总平面图
Site plan

3　中央庭院
Central Yard

4　会议室室内采用与主立面呼应的木色，体现萝岗区年轻、活力的城市性格
Interior wood color of the conference room echoes the main façade, embodying the urban vitality of Luogang District

3

1

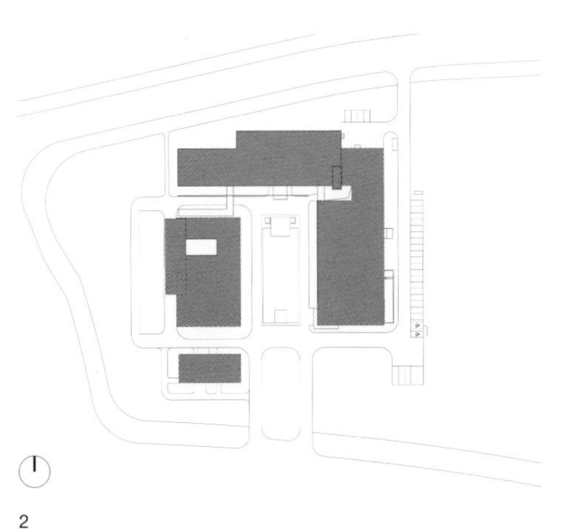

2

4

粤澳中医药科技产业园展示厅
Exhibition Hall, Guangdong-Macau Traditional Chinese Medicine Technology Industrial Park

2012-2013　珠海 横琴 / Hengqin Zhuhai

粤澳合作中医药产业园是广东省与澳门共同开发横琴岛的启动项目，具有重要的战略意义。本项目位于产业园的核心区域，是产业园区的第一个建成项目。它包含了展示中心及影院、VIP接待厅、投资者接待室、会议室、办公室等辅助用房等一系列功能。是对外展示产业园区发展蓝图，招商洽谈的重要平台。

展示厅形象清丽脱俗，宛如喧嚣都市里萌发的一片新叶。主展厅外立面采用铝板及玻璃材料，搭配了绿、白、黄等鲜亮色调，凸显建筑灵动活泼、奋发向上的形象，向人们传递充满活力与希望的信号。垂直的玻璃反射出阳光的灿烂、海空的湛蓝、绿意盎然。不同的材质拼贴与值得体味的建筑细部，创作出一种柔和的界面，弱化了建筑轮廓，模糊其与自然环境的界线，体现中医药天人合一、源于自然的哲理。

根据建筑的内部功能，建筑体现了明确的体量组合及虚实对比关系。其内部功能分区明确，流线清晰便捷。公共空间采用大面积玻璃围合，结合外部的休闲庭院，室内外空间相互渗透。展示馆是一座临时建筑，面对有限的投资资金，紧张的设计及建设周期，设计团队采用了简洁的形体，小中见大，注重细节，向业主及产业园区提交了一份满意的答卷。

As the kick-off project of Hengqin Island, the Traditional Chinese Medicine Technology Industrial Park jointly developed by Guangdong and Marco is of great strategic significance. Located in the core area of the Industrial Park, this firstly completed Project comprises exhibition center and cinema, VIP reception hall, investor reception room, conference room and offices.

The elegant exhibition hall appears like a fresh leat sprouted in the bustling metropolis. The façade of the main exhibition hall glazing in bright green, white and yellow communicates the vitality and hope. The assembling of different materials and impressive details contribute to a gentle interface, weakening the building profile and blurring the boundary between the building and natural surroundings, thus demonstrate the philosophy of Chinese traditional medicine, i.e. "unity between nature and human".

Functionally, the design shows clear volume combination and contrast between void and solid with clear functional zoning and convenient circulations. In view of the given budget and tight design and construction period for this temporary building, the design team presents a project with concise form and impressive details to satisfy the Client and Industrial Park.

项目地点：珠海市横琴中医药科技产业园区
建设时间：2012—2013年
用地面积：6245m²
建筑面积：2347m²
建筑层数：2层
建筑高度：12m

Location: Hengqin Traditional Chinese Medicine Technology Industrial Park, Zhuhai
Construction: 2012-2013
Site:6,245m²
GFA: 2,347m²
Number of floors: 2
Building height: 12m

1 总平面图
Site plan

2 建筑体量简洁、轻盈，虚实对比强烈，室内外空间相互融入
Simple and light building volume, strong solid-void, contrast, integration of inner and outer spaces

3 建筑造型洋溢着春意盎然的色彩
Vigorous architectural form

4 西立面
West Façade

5 南立面
South Façad

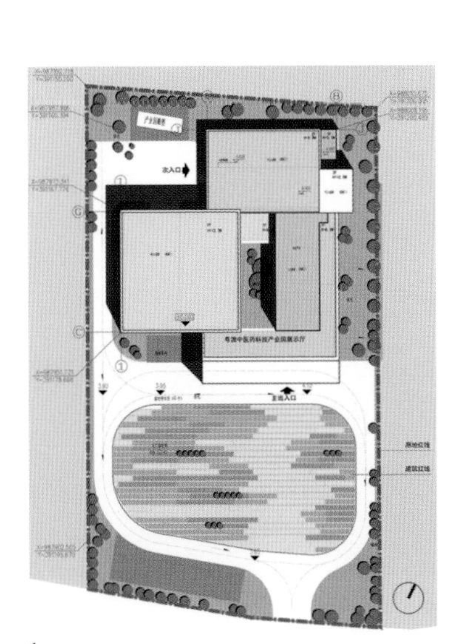

1

2

3

4

5

ADG·机场设计研究院办公楼
ADG Office Building

2008—2009　　广州 芳村 / Fangcun Guangzhou

信义会馆是一个由工业厂区改造而成的办公园区。它坐落于广州芳村白鹅潭畔，面朝珠江。茂密的老榕树、青砖路与红砖厂房互相映衬，形成了园区内独特的环境氛围。因此，一些创意产业，如广告公司、摄影工作室、画廊、设计公司等纷纷选择这里作为办公场地。

ADG·机场所的办公楼由一座红砖厂房与一栋新建的两层钢结构建筑组成。信义会馆的业主为我们提供了一个设计自己办公空间的机会，建筑及装修考虑了团队的现实需求与发展规划。老厂房内部的大空间完全被保留下来，为设计师提供了开放、宽敞的办公场所。厂房瓦顶下方满吊彩钢夹芯屋面板，改善了瓦屋顶隔热、漏雨、瓦片跌落等问题。新建的钢结构部分则提供独立办公室、会议室等小空间。大面积的落地玻璃窗把园区优美的景色全部纳入到内部的办公空间中。办公楼大堂通高两层，从入口处可以透过二层的玻璃走廊看到老厂房的坡屋面，形成独具特色的入口景观。精细的钢结构建筑与粗犷的红砖厂房巧妙地结合在一起，内部空间功能划分清晰，使用高效，从建筑形式、空间、色彩、材质等多方面体现了现代与传统的碰撞和融合。

这座办公楼是一个旧建筑活化使用的好案例。同时，它也承载着ADG·机场所团队"好设计、用心做"的经营理念，是团队的一张重要名片。

Xinyi Place is an office park transformed from an industrial zone. It is located at White Swan Waterfront, Fangcun, Guangzhou, facing the Pearl River. Leafy old banyans, grey brick roads and red brick plant buildings set one anther off nicely o form the unique atmosphere here.

ADG Office Building is composed of a red brick plant building and a new two-storey steel structure building. The generous space inside the old plant was retained to provide the designers with an open and spacious office. The colored steel sandwich panels are suspended below the tile roof to improve the thermal insulation and mitigate troubles caused by rain leakage or fallen roof tiles. The new steel structure is to create individual offices conference rooms, and other small closed spaces. The lobby is in two-floor height, while the sloping roof of the old plant is visible at the entrance through the glass corridor on the second floor and thus provides a peculiar view for the entrance. The delicate steel structure building and the rough red brick plant building form an interesting contrast which makes clear division between different internal spatial functions for efficient utilization of space.

This office building, as a name card, embodies and represents the business philosophy of the ADG team, that it, "to produce excellent designs with a passionate heart".

项目地点：广州芳村信义会馆
设计时间：2008—2009年
建设时间：2009年
建筑面积：$1273m^2$
建筑层数：2层

Location: Xinyi Place, Fangcun, Guangzhou
Design:2008-2009
Construction: 2009
GFA: ,$1273m^2$
Number of floors: 2

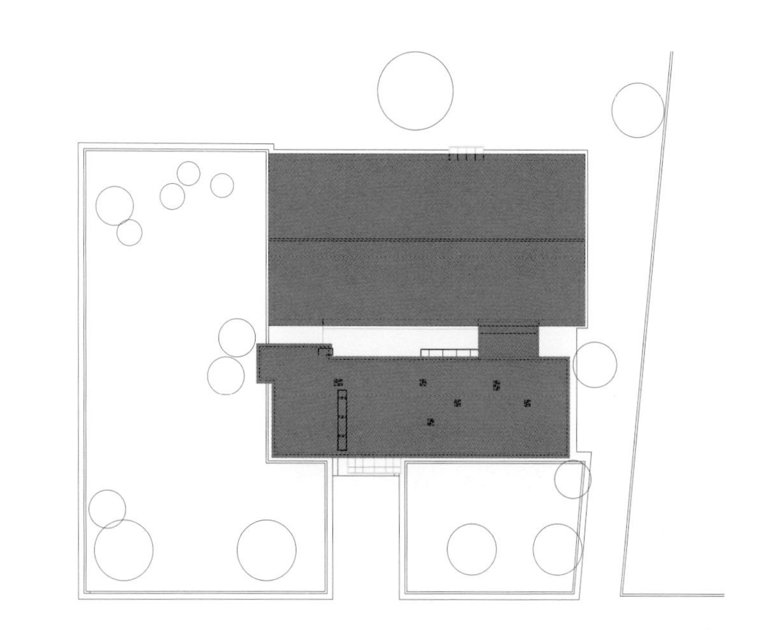

1

1　总平面图
Site plan

2　钢结构、玻璃、红砖等现代与传统材料调和结合，既体现建筑的现代性，也表达了对传统的尊重
The steel structure, glass and red bricks are blended together to represent the modernity of the building and show respect for the tradition

2

1

2

3

4

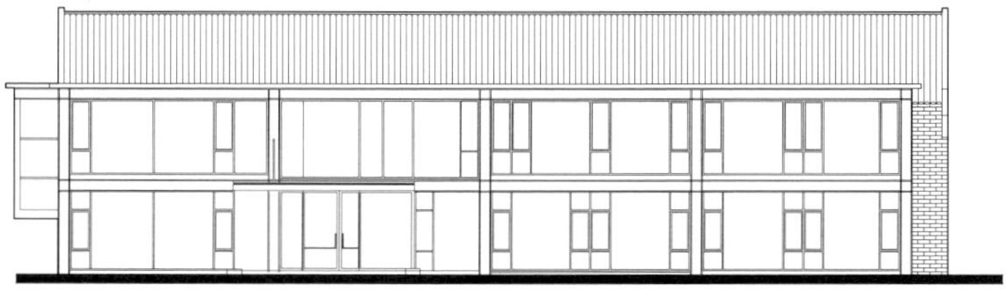

5

1 改造前的信义会馆办公楼，前身是一座红砖厂房
ADG Office Building used to be a red brick plant building before renovation

2 改造后的背立面，在原有建筑基础上增加通透玻璃
Back elevation after renovation features the added clear glazing

3 老厂房成为办公空间
The former plant building becomes office space

4 位于新旧建筑之间的二层露台花园，形成舒适的休憩空间
Terrace garden on F2 between the new and old building offers cozy leisure space

5 新与旧并存融合的建筑立面
Façade interweaving the old and the new

1

2

3

1 从入口门厅可以看到二层的玻璃廊道、花园的绿植及厂房坡屋面，形成独具特色的入口景观
The glass corridor on F2, green plants in the garden and the sloped roof of the plant building are visible from the entrance foyer, forming unique scenery at the entrance

2 从二层玻璃廊道看门厅及园区，青砖路与红砖厂房互相映衬，形成了园内独特的环境氛围
Viewed from the glass corridor on F2 into the foyer and park, the grey brick road and red brick plant building set off each other and foster distinctive environment in the park

3 视野通透的二层廊道，园外景色映入眼帘，与右侧的旧厂房并存共融
The corridor on F2 allows for unblocked view into the garden, thus harmoniously co-exists with the old plant buildings on the right

新增的钢结构正立面
Front elevation of new steel structure

Chronology 作品年表
Awards 获奖项目
Academic Achievements 学术成果
Afterword 后记

2004

广州新白云国际机场一期航站楼
项目地点：广东省广州市
建成时间：2004年
建筑面积：35.3万m²
Terminal 1, Guangzhou New Baiyun International Airport
Location: Guangzhou, Guangdong Province
Completion:2004
GFA: 353,000m²

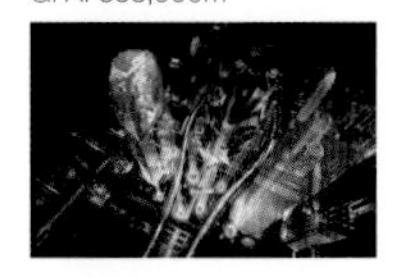

广州西塔
项目地点：广东省广州市
设计时间：2004年
建筑面积：49.29万m²
Guangzhou West Tower
Location: Guangzhou, Guangdong Province
Design:2004
GFA: 492,900 m²

广汽丰通钢业有限公司一期工程
项目地点：广东省广州市
设计时间：2004年
建筑面积：3.63万m²
Phase I Project of GAC TOYOTA Tsusho Steel Co., Ltd.
Location: Guangzhou, Guangdong Province
Design:2005
GFA: 36,300 m²

佛山恒福花园
项目地点：广州 佛山
设计时间：2004年
建筑面积：约45.14万m²
Foshan Hengfu Garden
Location: Foshan, Guangzhou
Foshan Hengfu Garden
Design:2004
GFA: approx. 451,400m²

杭州依江花园
项目地点：杭州 闻堰镇
设计时间：2004年
建筑面积：约43.73万m²
Hangzhou Yijiang Gardon
Location: Wenyan Town, Hangzhou
Design:2004
GFA: approx. 437,300m²

2005

广州新白云国际机场旅客航站楼东三、西三指廊及相关连接楼
项目地点：广东省广州市
建成时间：2009年
建筑面积：14.8万m²
East III, West III Pier and Connecting Building, Guangzhou New Baiyun International Airport
Location: Guangzhou, Guangdong Province
Completion:2009
GFA: 148,000 m²

广州新白云国际机场一期航站楼国际流程改造
项目地点：广东省广州市
设计时间：2009年
建筑面积： 2200m²
International Flow Improvement, Phase I of Guangzhou New Baiyun International Airport
Location: Guangzhou, Guangdong Province
Design: 2009
GFA:2200m²

东莞市商业中心区F区（海德广场）
项目地点：广东省东莞市
设计时间：2006年
建成时间：2013年
建筑面积：21.7万m²
Dongguan Commercial Center Zone F (Hyde Plaza)
Location: Dongguan, Guangdong Province
Design:2006
Completion:2013
GFA: 217,000m²

2006

广州洲头咀高级住宅项目
项目地点：广东省广州市
设计时间：2006年
建筑面积：21.25万m²
High-end Residential Project, Zhoutouzui, Guangzhou
Location: Guangzhou, Guangdong Province
Design:2006
GFA:212,500m²

惠州金山湖游泳跳水馆
项目地点：广东省惠州市
设计时间：2006年
建成时间：2010年
建筑面积：2.45万m²
Jinshan Lake Swimming and Diving Complex, Huizhou
Location: Huizhou, Guangdong Province
Design:2006
Completion:2010
GFA: 24,500m²

广州白云国际机场扩建工程-二号航站楼及配套设施
项目地点：广东省广州市
设计时间：2006-2010年（分离站坪概念）
2012至今 （北站坪概念）
建筑面积：63.4万平方米（2035年）
Expansion of Guangzhou New Baiyun International Airport - Terminal 2 and Supporting Facilities
Location: Guangzhou, Guangdong Province
Design: 2006-2010 (Separated apron concept)
2012 up to date (North apron concept)
GFA:634,000m² (2035)

昆明（长水）机场
项目地点：云南省昆明市
设计时间：2006年
建筑面积：29.3万m²（2035年）
Terminal of Kunming (Changshui) Airport
Location: Kunming, Yunnan Province
Design:2006
GFA:293,000m² (2035)

昆明巫家坝机场改造
项目地点：云南省昆明市
设计时间：2006年
建筑面积： 8.20万m²
Redevelopment for Kunming Wujiaba Airport
Location: Kunming, Yunnan Province
Design:2006
GFA:82,000m²

武汉火车站（武广高铁）
项目地点：湖北省武汉市
建成时间：2009年
建筑面积：10.68万m²
Wuhan Railway Station (Wuhan-Guangzhou High-speed Railway)
Location: Wuhan, Hubei Province
Completion: 2009
GFA:106,800m²

广州科学城科技人员公寓
项目地点：广东省广州市
建成时间：2010年
建筑面积：10.53万m²
Scientists' Apartment, Guangzhou Science City
Location: Guangzhou, Guangdong Province
Completion:2010
GFA: 105,300m²

2007

广州花都天马丽苑
项目地点：广东省广州市
设计时间：2007年
建筑面积：4.56万m²
Tian Ma Li Yuan Residential Development, Huadu District, Guangzhou
Location: Guangzhou, Guangdong Province
Design:2007
GFA:45,600m²

揭阳潮汕机场航站楼及配套工程
项目地点：广东省揭阳市
建成时间：2011年
建筑面积：5.87万m²
Jieyang Chaoshan Airport Terminal and Supporting Works
Location: Jieyang, Guangdong Province
Completion: 2011
GFA: 58,700m²

广州市城市规划展览中心
项目地点：广东省广州市
设计时间：2007年
建筑面积：5.94万m²
Guangzhou Urban Planning Exhibition Center
Location: Guangzhou, Guangdong Province
Design:2007
GFA:59,400 m²

海口海航机库
项目地点：海南省海口市
设计时间：2007年
建筑面积：2.01万m²
Hainan Airlines Aircraft Hangar,Haikou
Location:Haikou, Hainan Province
Design:2007
GFA:20,100 m²

南宁吴圩机场航站楼扩建工程
项目地点：广西壮族自治区南宁市
设计时间：2007年
建筑面积：6.28万m²
Terminal Building Expansion, Nanning Wuxu International Airport
Location:Nanning, Guangxi Zhuang Autonomous Region
Design:2007
GFA:62,800 m²

广州亚运村媒体村居住区
设计地点：广东省广州市
设计时间：2007年
建筑面积：41.37万m²
Residential Zone, Media Village of Asian Games Town, Guangzhou
Location: Guangzhou, Guangdong Province
Design:2007

GFA:413,700m^2

广州亚运馆
项目地点：广东省广州市
建成时间：2010年
建筑面积： 6.53万m^2
Guangzhou Asian Games Gymnasium
Location: Guangzhou, Guangdong Province
Completion: 2010
GFA: 65,300m^2

广州新城亚运主媒体中心
项目地点：广东省广州市
建成时间：2010年
建筑面积：6.00万m^2
Main Media Center, Asian Games Town, Guangzhou
Location: Guangzhou, Guangdong Province
Completion:2010
GFA: 60,000 m^2

广州亚运会省属场馆游泳跳水馆
项目地点：广东省广州市
设计时间：2008年
建筑面积：3.36万m^2
Natatorium and Diving Stadium of Guangdong Province, Guangzhou Asian Games
Location: Guangzhou, Guangdong Province
Design:2008
GFA:33,600m^2

长沙黄花国际机场扩建工程
项目地点：湖南省长沙市
设计时间：2007年
建筑面积：35.3万m^2
Design Competition For New Terminal Building Of Huanghua International Airport,Changsha
Location:Changsha, Hunan Province
Design: 2008
GFA:353,000m^2

2008

南宁竹溪路住宅项目
项目地点：广西壮族自治区南宁市
设计时间：2008年
建筑面积：43.37万m^2
Residential Project, Zhu Xi Lu, Nanning
Location:Nanning, Guangxi Zhuang Autonomous Region
Design:2008
GFA:433,700m^2

揭阳潮汕民用机场工作区规划及单体设计
项目地点：广东省揭阳市
设计时间：2008年
建筑面积：12.02万m^2
Planning for Working Area and Singular Building Design of Jieyang Chaoshan Civil Airport
Location:Jieyang, Guangdong Province
Design:2008
GFA:120,200m^2

桂林两江国际机场航站楼扩建工程
项目地点：广西壮族自治区桂林市
设计时间：2008年
建筑面积： 5.31万m^2
Expansion of Terminal Building, Guilin Liangjiang International Airport
Location:Guililn, Guangxi Zhuang Autonomous Region
Design:2008
GFA:53,100m^2

广州花都区东风体育馆
项目地点：广东省广州市
建成时间：2010年
建筑面积：3.14万m^2
Dongfeng Gymnasium, Huadu District, Guangzhou
Location: Guangzhou, Guangdong Province
Completion:2010
GFA: 31,400m^2

凯旋华美达酒店改扩建方案竞赛
项目地点：广东省广州市
设计时间：2008年
建筑面积：8.85万m^2
Competition for Expansion and Renovation of Ramada Pearl Hotel
Location: Guangzhou, Guangdong Province
Design:2008
GFA:88,500m^2

ADG·机场设计研究院办公楼
项目地点：广东省广州市
建成时间：2009年
建筑面积： 1,237m^2
ADG Office Building
Location: Guangzhou, Guangdong Province
Completion:2009
GFA: 1,237m^2

2009

北京回龙观公寓
项目地点：北京市
设计时间：2009年
建筑面积：3.94万m^2
Hui Long Guan Apartment, Beijing
Location: Beijing
Design:2009
GFA:39,400m^2

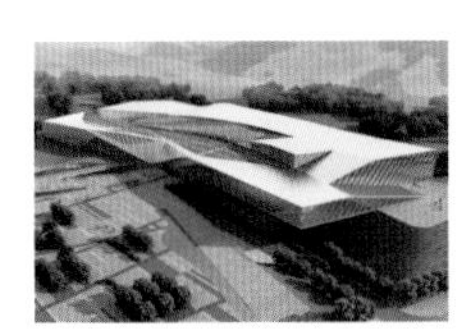

广西城市规划建设展示馆
项目地点：广西壮族自治区南宁市
设计时间：2009年
建筑面积：4.39万m^2
Guangxi Urban Planning and Construction Exhibition Museum
Location:Nanning, Guangxi Zhuang Autonomous Region
Design:2009
GFA:43,900 m^2

北海冠岭五星酒店
项目地点：北海冠岭滨海景区
设计时间：2009年
建筑面积：3.5万m^2
The five-star hotel Guanling ,Beihai
Location:The coastal area of Guanling ,Beihai
Design:2009
GFA:35,000 m^2

广州气象卫星地面站B站区
项目地点：广东省广州市
建成时间：2013年
建筑面积：5539m^2
Business Building of Guangzhou Meteorological Satellite Ground Station Zone B
Location: Guangzhou, Guangdong Province
Completion:2013
GFA:5,539 m^2

南宁高新区总部基地三期综合楼
项目地点：广西壮族自治区南宁市
设计时间：2009年
建筑面积：12.42万m^2
Complex, Phase III, Headquarter of Nanning Hi-Tech Zone
Location:Nanning, Guangxi Zhuang Autonomous Region
Design:2009
GFA:124,200 m^2

南宁吴圩国际机场新航站区及配套设施扩建工程
项目地点：广西壮族自治区南宁市
设计时间：2009年
建筑面积：12.97万m^2（2020）
Design Competition for Terminal Building and Expansion of New Terminal Area and Supporting Facilities, Wuxu International Airport, Nanning
Location:Nanning, Guangxi Zhuang Autonomous Region
Design:2009
GFA:129,700m^2 (2020)

三联·金洲号
项目地点：广东省惠州市
设计时间：2009年
建筑面积：35.00万m^2
Sanlian-Jinzhou
Location: Huizhou, Guangdong Province
Design:2009
GFA:350,000m^2

深圳机场新航站区地面交通中心（GTC）
项目地点：深圳市
建成时间：2012年
建筑面积：5.81万m^2
Ground Transportation Center (GTC) of New Terminal Area, Shenzhen Airport
Location: Shenzhen, Guangdong Province
Completion:2012
GFA:58,100m^2

武汉天河机场三号航站楼
项目地点：湖北省武汉
设计时间：2009年
建筑面积：12.12万m^2
Terminal 3, Wuhan Tianhe Airport
Location: Wuhan, Hubei Province
Design:2009
GFA:121,200m^2

亚运会形象设计
项目地点：广东省广州市
设计时间：2009年
mage Design for Guangzhou Asian Games
Location: Guangzhou, Guangdong Province
Design:2009

广州新白云国际机场一号航站楼亚运三大流程改造
项目地点：广东省广州市
设计时间：2009年
建筑面积： 1.87万m^2
Three Operation Flow Improvement Works for Asian Games, Terminal 1, Guangzhou New Baiyun International Airport
Location: Guangzhou, Guangdong Province
Design:2009
GFA:18,700m^2

中国医药医疗器械展示中心
项目地点：广东省广州市
设计时间：2009年
建筑面积：19.93万m^2
Chia Medical Devices Exhibition Center
Location: Guangzhou, Guangdong Province
Design:2009
GFA:199,300m^2

重庆江北国际机场东航站区及配套设施
项目地点：重庆市
设计时间：2009年
建筑面积：32.78m²
East Terminal Area and Supporting Facilities, Chongqing Jiangbei International Airport
Location: Chongqing
Design:2009
GFA:327,800m²

2010

LED中国九州城
项目地点：广东省中山市
设计时间：2010年
建筑面积：3.74万m²
LED Jiuzhou City
Location: Zhongshan, Guangdong Province
Design:2010
GFA:37,400m²

柳州市马鹿山公园东侧城中村改造
项目地点：广西壮族自治区柳州市
设计时间：2010年
建筑面积：26.72万m²
Redevelopment of Urban Village on the east of Ma Lu Shan Park, Liuzhou
Location:Liuzhou, Guangxi Zhuang Autonomous Region
Design:2010
GFA:267,200m²

广州海珠区龙潭村三旧改造项目概念方案
项目地点：广州 龙潭村
设计时间：2010年
研究范围：约121.33hm²
Conceptual Scheme for Longtan Village Redevelopment Project, Haizhu District, Guangzhou
Location:Longtan Village, Guangzhou
Design:2010
Study scope: approx.121.33ha

柳州市三淋村片区回建房小区
项目地点：广西壮族自治区柳州市
设计时间：2010年
建筑面积：44.61万m²
Resettlement Quarter, Sanlin Village, Liuzhou
Location:Liuzhou, Guangxi Zhuang Autonomous Region
Design:2010
GFA:446,100m²

蒙东电力公司呼和浩特基地
项目地点：内蒙古自治区呼和浩特市
设计时间：2010年
建筑面积：14.30万m²
Hohhot Campus of East Inner Mongolia Electrical Power Company Limited
Location: Hohhot, Inner Mongolia Autonomous Region
Design:2010
GFA:143,000m²

南宁德利·东盟国际文化广场
项目地点：广西壮族自治区南宁市
设计时间：2010年
建筑面积：36.9万m²
AICC (ASEAN International Culture Plaza), Nanning
Location:Nanning, Guangxi Zhuang Autonomous Region
Design:2010
GFA:369,000m²

南宁天龙财富中心
项目地点：广西壮族自治区南宁市
设计时间：2010年
建筑面积：30.33万m²
Tianlong Fortune Center, Nanning
Location:Nanning, Guangxi Zhuang Autonomous Region
Design:2010
GFA:303,300m²

深圳机场航站区扩建工程停车场及配套商务区项目
项目地点：广东省深圳市
设计时间：2010年
建筑面积：26.40万m²
Parking and Supporting Business District, Expansion of Terminal Area, Shenzhen Airport
Location: Shenzhen, Guangdong Province
Design:2010
GFA:264,000m²

广州白云机场南航GAMECO飞机维修设施二期工程
项目地点：广东省广州市
设计时间：2010年
建筑面积：4.78万m²
GAMECO Aircraft Maintenance Facilities (Phase II), China Southern Airlines, Guangzhou Baiyun International Airport
Location: Guangzhou, Guangdong Province
Design:2010
GFA:47,800m²

中山小榄金融大厦
项目地点：广东省中山市
设计时间：2010年
建筑面积：8.74万m²
Zhongshan Xiaolan Financial Plaza
Location: Zhongshan, Guangdong Province
Design:2010
GFA:87,400m²

中新（广州）知识城规划展示厅
项目地点：广东省广州市
设计时间：2010年
建筑面积：1500m²
Exhibition Hall, Sino-Singapore Guangzhou Knowledge City
Location: Guangzhou, Guangdong Province
Design:2010
GFA:1,500m²

2011

汕头柏嘉半岛（二期）
项目地点：广东省汕头市
设计时间：2011年
建筑面积：51.93万m²
Bai Jia Ban Dao Residential Development, Shantou (Phase II)
Location: Shantou, Guangdong Province
Design:2011
GFA: 519,300m²

东莞勤上LED照明研发设计中心
项目地点：广东省东莞市
设计时间：2011年
建筑面积：8.12万m²
Kingsun LED Lighting R&D Center, Dongguan
Location: Dongguan, Guangdong Province
Design:2011
GFA:81,200m²

佛冈汤塘镇鹤鸣洲温泉度假村
项目地点：广东省清远市
设计时间：2012年
建筑面积：10.80万m²
He Ming Zhou Hotspring Resort
Location: Qingyuan, Guangdong Province
Design:2012
GFA:108,000m²

广西扶绥金源财富广场
项目地点：广西壮族自治区崇左市
设计时间：2011年
建筑面积：43.02万m²
Jinyuan Fortune Center, Fusui, Guangxi Province
Location:Chongzuo, Guangxi Zhuang Autonomous Region
Design:2011
GFA:430,200m²

海丰爱琴湾旅游创意园区规划
项目地点：广东省汕尾市
设计时间：2011年
规划面积：276.7万m²
Planning for Aiqin Bay Tourism Creative Park, Haifeng
Location: Shanwei, Guangdong Province
Design:2011
Planning area: 2,767,000 m²

鹤山海德庄园
项目地点：广东省鹤山市
设计时间：2011年
建筑面积：59.49万m²
Hyde Manor, Heshan
Location: Heshan, Guangdong Province
Design:2011
GFA:594,900m²

罗浮山悦榕庄酒店
项目地点：广东省惠州市
设计时间：2011年
建筑面积：5.52万m²
Banyan Tree Hotel, Mt. Luofu
Location: Huizhou, Guangdong Province
Design:2011
GFA:55,200m²

香港新福港地产·佛山新福港广场
项目地点：广东省佛山市
设计时间：2011年
建筑面积：55.07万m²
SFK · SFK Plaza, Foshan
Location: Foushan, Guangdong Province
Design:2011
GFA:550,700m²

烟台潮水机场航站楼
项目地点：山东省烟台市
设计时间：2011年
建筑面积：7.91万m²
Design Competition for New Terminal Building of Yantai Tidewater International Airport
Location: Yantai, Shandong Province
Design:2011
GFA:79,100m²

中新广州知识城公共服务中心
项目地点：广东省广州市
设计时间：2011年
建筑面积：0.82万m²
Public Service Center, Sino-Singapore Guangzhou Knowledge City
Location: Guangzhou, Guangdong Province
Design:2011
GFA:8,200m²

中新（广州）知识城展厅配套接待中心
项目地点：广东省广州市
设计时间：2011年
建筑面积：7885m²
Exhibition Hall, Sino-Singapore Guangzhou Knowledge City
Location: Guangzhou, Guangdong Province
Design:2011
GFA:7,885m²

保利地产·珠海横琴发展大厦
项目地点：广东省珠海市
设计时间：2011年
建筑面积：约23万m²
Poly · Zhuhai Hengqin Development Building
Location: Zhuhai, Guangdong Province
Design:2011
GFA:about 230,000m²

香港新福港地产·广州鼎峰
项目地点：广东省广州市
设计时间：2011年
建筑面积：约42万m²
SFK · DF Project, Guangzhout

Location:Guangzhou, Guangdong Province
Design:2011
GFA:420,000m^2

珠海横琴新区横琴发展大厦建筑方案设计及周边地块城市设计
项目地点：珠海横琴
设计时间：2011年
设计范围：43.9万m^2
Architectural Schematic Design for Hengqin Development Mansion, Hengqin District, Zhuhai and Urban Design for Surrounding Plots
Location: zhuhai hengqin
Design: 2011
Urban design scope: 439,000m^2

2012

南宁北湖劲源综合楼
项目地点：广西壮族自治区南宁市
设计时间：2012年
建筑面积：11.02万m^2
General Building of Bei Hu Jin Yuan Residential Development, Nanning
Location:Nanning, Guangxi Zhuang Autonomous Region
Design:2012
GFA:110,200m^2

广州白云机场公安应急指挥中心
项目地点：广东省广州市
设计时间：2012年
建筑面积：1.55万m^2
Emergency Command Center, Guangzhou Baiyun International Airport
Location: Guangzhou, Guangdong Province
Design:2012
GFA:15,500m^2

海口美兰国际机场二号航站楼
项目地点：海南省海口市
设计时间：2012年
建筑面积：25.24万m^2
Terminal 2, Haikou Meilan International Airport
Location:Haikou, Hainan Province
Design:2012
GFA:252,400m^2

粤澳中医药科技产业园展示厅
项目地点：珠海 横琴
建成时间：2013年
建筑面积：2347m^2
Exhibition Hall, Guangdong-Macau Traditional Chinese Medicine Technology Industrial Park
Location: Guangzhou, Guangdong Province
Completion:2013
GFA: 2,347m^2

香港新福港地产·鹤山峻廷湾二期
项目地点：广东省鹤山市
设计时间：2012年
建筑面积：25.95万m^2
SFK ·Jun Ting Wan (Phase II), Heshan
Location: Heshan, Guangdong Province
Design:2012
GFA:259,500m^2

丽江玉龙县无量寿生态旅游小镇
项目地点：云南省丽江市
设计时间：2012年
建筑面积：12.4万m^2
Infinite Life Ecological Tourist Town, Yulong County, Lijiang
Location: Lijiang, Yunnan Province
Design:2012
GFA:124,000m^2

佛山欧浦国际商业中心
项目地点：广东省佛山市
设计时间：2012年
建筑面积：22.2万m^2
Europol International Commercial Center, Foshan
Location: Foushan, Guangdong Province
Design:2012
GFA:222,000m^2

珠海横琴地方税务局新办公楼
项目地点：广东省珠海市
设计时间：2012年
建筑面积：1.2万m^2
New Office Building of Zhuhai Municipal Local Taxation Bureau
Location: Zhuhai, Guangdong Province
Design:2012
GFA:12,000m^2

南方广播影视创意基地一期
项目地点：广东省广州市
设计时间：2012年
建筑面积：54.4万m^2
Creative Base (Phase I), Southern Media Corporation
Location: Guangzhou, Guangdong Province
Design:2012
GFA:544,000m^2

2013

岭南V谷—广州国际智能科技园
项目地点：广东省广州市
设计时间：2013年
建筑面积：54.17万m^2
Lingnan Valley – Planning for Guangzhou International Intelligent Science Park
Location: Guangzhou, Guangdong Province
Design:2013
GFA:541,700m^2

广州新白云国际机场一号航站楼东连接楼连廊建设工程
项目地点：广东省广州市
设计时间：2013年
建筑面积：1.21万m^2
Sky Walk of East Connecting Building, Terminal 1, Guangzhou New Baiyun International Airport
Location: Guangzhou, Guangdong Province
Design:2013
GFA:12,100 m^2

广州南沙新区蕉门河中心区城市设计暨建筑方案
项目地点：广东省广州市
设计时间：2013年
建筑面积：98.00万m^2
Urban Design and Architectural Design for Jiaomen River Central Area of Nansha, Guangzhou
Location: Guangzhou, Guangdong Province
Design:2013
GFA:980,000 m^2

三亚凤凰路与迎宾路交界西北侧项目
项目地点：海南省三亚市
设计时间：2013年
建筑面积：7.69万m^2
Project to North-west of Intersection of Feng Huang Lu and Ying Bin Lu, Sanya
Location:Sanya, Hainan Province
Design:2013
GFA:76,900m^2

上海浦东国际机场南航站区卫星厅
项目地点：上海市
设计时间：2013年
建筑面积：52.90万m^2
Satellite Concourse, South Terminal Area, Shanghai Pudong International Airport
Location: Shanghai
Design:2013
GFA:529,000m^2

宜昌奥林匹克体育中心概念规划及建筑设计
项目地点：湖北省宜昌市
设计时间：2013年
建筑面积：25.4万m^2
Conceptual Planning and Architectural Design Competition for Yichang Olympic Sports Center
Location: Yichang, Hubei Province
Design:2013
GFA:254,000m^2

广州开发区知祥公寓
项目地点：广东省广州市
设计时间：2013年
建筑面积：6.99万m^2
Zhixiang Apartment, Guangzhou Development Zone
Location: Guangzhou, Guangdong Province
Design:2013
GFA:69,900m^2

从化市民之家投标方案及深化方案
项目地点：广东省从化市
设计时间：2013年
建筑面积：24万m^2
Conghua New City-Citizen Home
Location: Coonghua, Guangdong Province
Design:2013
GFA:240,000m^2

保利地产·珠海横琴保利国际广场
项目地点：珠海横琴
设计时间：2013
建筑面积：10.24万m^2
Poly·Zhuhai Hengqin Poly International Plaza
Location:zhuhai hengqin
Design:2013
GFA:102,400m^2

2014

中海广钢新城
项目地点：广东省广州市
设计时间：2014年
建筑面积：19.46万m^2
Zhonghai Guangzhou Iron & Steel New City
Location: Guangzhou, Guangdong Province
Design:2014
GFA:194,600m^2

万威森林公园二期
项目地点：广东省珠海市
设计时间：2014年
建筑面积：50.97万m^2
Wanwei Forest Park (Phase II)
Location: Zhuhai, Guangdong Province
Design:2014
GFA:509,700m^2

珠海斗门总部大厦、展览中心及市民公园
项目地点：广东省珠海市
设计时间：2014年
建筑面积：3.63万m^2
Doumen Headquarter Building, Exhibition Center and Citizen Park, Zhuhai
Location: Zhuhai, Guangdong Province
Design:2014
GFA:36,300m^2

岭南V谷—广州国际智能科技园建筑方案投标
项目地点：广东省广州
设计时间：2014年
建筑面积：54.16万m^2
Lingnan Valley – Architectural Bid Proposal of Guangzhou International Intelligent Science Park
Location: Guangzhou, Guangdong Province
Design: 2014
GFA: 541,600m^2

2004

广州西塔
国际竞赛优胜方案

2005

广州新白云国际机场一期航站楼
第五届詹天佑土木工程大奖
"全国十大建设科技成就"称号
首届全国绿色建筑创新奖
广东省优秀工程设计一等奖

东莞市商业中心区F区（海德广场）
邀请竞赛中标

2006

广州白云国际机场迁建工程
2006年度全国优秀工程设计金奖

广州科学城科技人员公寓
国际竞赛中标

惠州金山湖游泳跳水馆
全国竞赛中标

2007

广州新白云国际机场一期航站楼
广东省优秀工程技术创新奖

长沙黄花国际机场扩建工程
广东省注册建筑师协会优秀建筑创作奖
国际竞赛第二名

揭阳潮汕机场航站楼及配套工程
国际竞赛中标

2008

广州亚运馆
国际竞赛中标

广州亚运主媒体中心
国际竞赛中标

广州花都区东风体育馆
全国竞赛中标

2009

广州亚运馆
住房和城乡建设部绿色建筑与低能耗建筑"双百"示范工程

重庆江北机场
广东省注册建筑师优秀建筑佳作奖

2010

广州新白云国际机场一期航站楼
中国建筑学会建筑创作大奖

广州亚运馆
2010年度china-designer中国室内设计年度评选优秀建筑空间设计金堂奖

武汉火车站
铁路优质工程勘察设计一等奖

2011

广州新白云国际机场一期航站楼
中国百年百项杰出土木工程（1912-2011）

广州亚运馆
中国百年百项杰出土木工程（1912-2011）
第十届詹天佑土木工程大奖
第十届中国土木工程詹天佑奖创新集体奖
广东省优秀工程设计奖一等奖
全国优秀工程勘察设计行业建筑工程设计一等奖
广东省注册建筑师协会优秀建筑创作奖
中国钢结构金奖
广东钢结构金奖"粤钢奖"设计一等奖

广州花都区东风体育馆
全国优秀工程勘察设计行业奖三等奖
广东省优秀工程设计二等奖
广东省注册建筑师优秀建筑佳作奖
广东钢结构金奖"粤钢奖"
第十届届中国室内设计大奖赛学会奖

惠州金山湖游泳跳水馆
全国优秀工程勘察设计行业二等奖
广东省优秀工程设计二等奖
广东省注册建筑师优秀建筑佳作奖
惠州市优秀工程设计一等奖

武汉火车站
第十届詹天佑土木工程奖
中国百年百项杰出土木工程（1912-2011）

广州亚运主媒体中心
广东省优秀工程设计三等奖

广州新白云国际机场东三西三指廊及相关连接楼
广东省优秀工程设计二等奖

广州气象卫星地面站B站区
广东省注册建筑师协会优秀建筑创作奖

广州科学城科技人员公寓
广东省优秀工程设计二等奖
广东省注册建筑师协会优秀建筑创作奖

广西城市规划建设展示馆
广东省注册建筑师协会优秀建筑创作奖

南宁吴圩国际机场新航站区及配套设施扩建工程
广东省注册建筑师协会优秀建筑佳作奖

广州市气象监测预警中心方案
广东省注册建筑师协会优秀建筑创作奖

保利地产·珠海横琴发展大厦
国际竞赛中标

珠海横琴新区横琴发展大厦建筑方案设计及周边地块城市设计
国际竞赛中标

2012

广州亚运馆
第六届中国建筑学会建筑创作优秀奖

惠州金山湖游泳跳水馆
第六届中国建筑学会建筑创作佳作奖

广州科学城科技人员公寓
第六届中国建筑学会建筑创作优秀奖

武汉火车站
芝加哥雅典娜建筑设计博物馆颁发2012年国际建筑奖

东莞LED照明研发中心
东莞市优秀建筑工程设计方案一等奖

2013

广州亚运馆
香港建筑师学会两岸四地建筑设计大奖优异奖

揭阳潮汕机场航站楼及配套工程
全国优秀工程勘察设计行业三等奖
广东省优秀工程勘察设计二等奖

珠海横琴新区横琴发展大厦建筑方案设计及周边地块城市设计
广东省注册建筑师协会优秀建筑佳作奖

南方广播影视创意基地一期
广东省注册建筑师协会优秀建筑佳作奖

从化新市民之家
全国竞赛中标

2014

广州亚运馆
AAA2014亚洲建筑师协会奖荣誉奖

保利地产·珠海横琴发展大厦
广东省注册建筑师协会优秀建筑佳作奖

2004

Guangzhou West Tower
The First Place of international competition

2005

Terminal 1, Guangzhou New Baiyun International
The 5th Tien-Yow Jeme Civil Engineering Prize
Top 10 National Construction Technology Achievement Award
The First "National Green Building Innovation Award"
The First Prize for Excellent Engineering Design Award of Guangdong Province

Dongguan Commercial Center Zone F (Hyde Plaza)
The First Place of national competition

2006

Terminal 1, Guangzhou New Baiyun International Airport
Golden Prize for National Excellent Engineering Exploration and Design Award

Scientists' Apartment, Guangzhou Science City
The First Place of international competition

Jinshan Lake Swimming and Diving Complex, Huizhou
The First Place of national competition

2007

Terminal 1, Guangzhou New Baiyun International Airport
Excellent Engineering Technology Innovation Award of Guangdong Province

Design Competition for New Terminal Building of Huanghua International Airport, Changsha
The Second Place of international competition
Excellent Architecture Creation Award by Guangdong Chapter of Association of Chinese Registered Architects

Jieyang Chaoshan Airport Terminal and Supporting Works
The First Place of international competition

2008

Guangzhou Asian Games Gymnasium
The First Place of international competition

Main Media Center, Asian Games Town, Guangzhou
The First Place of international competition

Dongfeng Gymnasium, Huadu District, Guangzhou
The First Place of national competition

2009

Guangzhou Asian Games Gymnasium
Model Project for Two 100-Top Green Building and Energy Saving by Ministry of Housing and Urban Rural Development of PRC

Chongqing Jiangbei Airport
Excellent Architecture Creation Award by Guangdong Chapter of Association of Chinese Registered Architects

2010

Terminal 1, Guangzhou New Baiyun International Airport
ASC Architectural Creation Award

Guangzhou Asian Games Gymnasium
Jin Tang Prize for China Interior Design Awards 2010 – Excellent Public Space Design

Wuhan Railway Station
The First Prize for Excellent Railway Engineering Exploration and Design

2011

Terminal 1, Guangzhou New Baiyun International Airport
"100 Outstanding Civil Engineering Projects from 1900 to 2010"

Guangzhou Asian Games Gymnasium
The First Prize for National Excellent Engineering Exploration and Design - Engineering Exploration and Design
"100 Outstanding Civil Engineering Projects from 1900 to 2010"
The First Prize for Excellent Engineering Design Award of Guangdong Province
Excellent Architecture Creation Award of Guangdong Chapter of Association of Chinese Registered Architects (ACRAGD)
The First Prize for Gold Prize for Steel Structure by Guangdong Provincial Society for Spatial Structures
The 10th Tien-Yow Jeme Civil Engineering Prize
Tien-yow Jeme Civil Engineering Prize (Innovation Team Prize)
Gold Prize for Steel Structure by China Construction Metal Structure Association

Dongfeng Gymnasium, Huadu District, Guangzhou
The Third Prize for National Excellent Engineering Exploration and Design
The 6th Excellent Architecture Creation Award by Guangdong Chapter of Association of Chinese Registered Architects
The Second Prize for Excellent Engineering Design Award of Guangdong Province
Gold Prize for Steel Structure of Guangdong Province (Yue Gang Award)
Society Prize for the 14th China Interior Design Awards

Jinshan Lake Swimming and Diving Complex, Huizhou
The Second Prize for National Excellent Engineering Exploration and Design
Excellent Architecture Creation Award by Guangdong Chapter of Association of Chinese Registered Architects
The Second Prize for Excellent Engineering Design Award of Guangdong Province
The First Prize for Excellent Engineering Designs of Huizhou

Wuhan Railway Station
The 10th Tien-Yow Jeme Civil Engineering Prize
"100 Outstanding Civil Engineering Projects from 1900 to 2010"

Main Media Center, Asian Games Town, Guangzhou
The Third Prize for Excellent Engineering Design Award of Guangdong Province

East III, West III Pier and relevant Connecting Building, Guangzhou New Baiyun International Airport
The Second Prize for Excellent Engineering Design Award of Guangdong Province

Business Building of Guangzhou Meteorological Satellite Ground Station Zone B
Excellent Architecture Creation Award by Guangdong Chapter of Association of Chinese Registered Architects

Scientists' Apartment, Guangzhou Science City
The Second Prize for Excellent Engineering Design Award of Guangdong Province
Excellent Architecture Creation Award by Guangdong Chapter of Association of Chinese Registered Architects

Guangxi Urban Planning and Construction Exhibition Hall
Excellent Architecture Creation Award by Guangdong Chapter of Association of Chinese Registered Architects

Design Competition for Terminal Building as Expansion of New Terminal Area and Supporting Facilities, Wuxu International Airport, Nanning
Excellent Architecture Creation Award of Guangdong Chapter of Association of Chinese Registered Architects

Meteorological Monitor & Alert Center,Guangzhou
Excellent Architecture Creation Award of Guangdong Chapter of Association of Chinese Registered Architects

Poly · Zhuhai Hengqin Development Building
The First Place of international competition

Architectural Schematic Design for Hengqin Development Mansion, Hengqin District, Zhuhai and Urban Design for Surrounding Plots
The First Place of international competition

2012

Guangzhou Asian Games Gymnasium
Excellent Award of the 6th ASC Architectural Creation Award

Jinshan Lake Swimming and Diving Complex, Huizhou
The 6th ASC Architectural Creation Award

Scientists' Apartment, Guangzhou Science City
Excellent Award of the 6th ASC Architectural Creation Award

Wuhan Railway Station
The International Architecture Award for 2012 by Museum of Architecture and Design, Chicago Athenaeum

Dongguan LED Lighting R&D Center
The First Prize for Excellent Architecture Design Award of Dongguan

2013

Guangzhou Asian Games Gymnasium
Merit Award, Cross-Strait Architectural Design Awards 2013 by Hong Kong Institute of Architects

Jieyang Chaoshan Airport Terminal and Supporting Works
The Third Prize for National Excellent Engineering Exploration and Design Award
The Second Prize for Excellent Engineering Exploration and Design of Guangdong Province

Architectural Schematic Design for Hengqin Development Mansion, Hengqin District, Zhuhai and Urban Design for Surrounding Plots
Excellent Architecture Creation Award by Guangdong Chapter of Association of Chinese Registered Architects

Creative Base (Phase I), Southern Media Corporation
Excellent Architecture Creation Award by Guangdong Chapter of Association of Chinese Registered Architects

Conghua New Citizen Home
The First Place of national competition

2014

Guangzhou Asian Games Gymnasium
The ARCASIA Award (AAA2014): Honor Award – Architecture

Poly · Zhuhai Hengqin Development Mansion
Excellent Architecture Creation Award by Guangdong Chapter of Association of Chinese Registered Architects

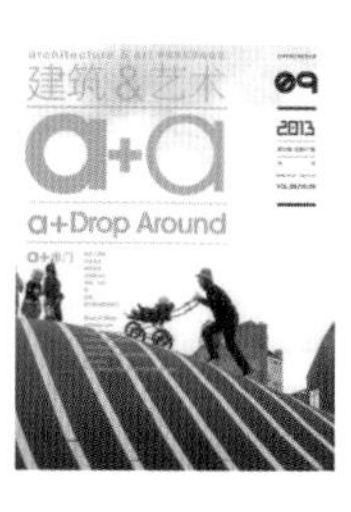

Papers of Architecture 建筑专业论文

干线机场航站楼创新实践——潮汕机场航站楼设计
作者：陈雄 潘勇
出版日期：2014年 《建筑学报》 2014年02期
Innovative Practice of Terminals in Major Airports – Design of Chaoshan Airport Terminal
Author: Chen Xiong, Pan Yong
Publication: *Architectural Journal*, Issue 02, 2014

游走在传统与现代之间
作者：陈雄
出版日期：2013年 《a+a》杂志 2013年10期
Between Tradition and Modernity
Author: Chen Xiong
Publication: *a+a*, Issue 10, 2013

好设计 用心做
作者：陈雄
出版日期：2013年 《a+a》杂志 2013年09期
Design with Heart and Soul
Author: Chen Xiong
Publication: *a+a*, Issue 09, 2013

一个不具备共性特征的住宅设计
作者：谢少明、郭胜
出版日期：2011年 《建筑学报》2011年第02期
A Housing Design without Common Features
Author: Xie Shaoming, Guo Sheng
Publication: *Architectural Journal*, Issue 02, 2011

广州科技人员公寓
作者：日本佐藤综合计画株式会社+ADG团队
出版日期：2011年 日本《新建筑》2011年8期
Scientists Apartment of Guangzhou Development District
Author: AXS SATOW INC. + ADG
Publication: *JA+U*, Issue 8, 2011

体育建筑的理性演绎——广州市花都东风体育馆
作者：郭胜、陈超敏
出版日期：2011年 《建筑学报》 2011年09期
Rational Deduction of Sports Buildings – Huadu Dongfeng Gymnasium, Guangzhou City
Author: Guo Sheng, Chen Chaomin
Publication: *Architectural Journal*, Issue 09, 2011

广州亚运馆
作者：潘勇、陈雄
出版日期：2010 《建筑学报》第10期
Guangzhou Asian Games Gymnasium
Author: Pan Yong, Chen Xiong
Publication: *Architectural Journal*, Issue 10, 2010

广州亚运馆设计与思考
作者：潘勇 陈雄
出版日期：2010 《建筑创作》第137期
Design and Reflection of Guangzhou Asian Games Gymnasium
Author: Pan Yong, Chen Xiong
Publication: *Architecture Creation*, Issue 137, 2010

机场航站楼发展趋势与设计研究
作者：陈雄
出版日期：2008 《建筑学报》第5期
Development Tendency and Design Research of Airport Terminals
Author: Chen Xiong
Publication: *Architectural Journal*,, Issue 5, 2008

机场航站楼设计的地域性思考——潮汕机场航站楼设计
作者：陈雄、潘勇
出版日期：2008 《南方建筑》第1期
Regionalism of Airport Terminal Design - Chaoshan Airport Terminal
Author: Chen Xiong, Pan Yong
Publication: *South Architecture*, Issue 1, 2008

构筑实现梦想的舞台——惠州市金山湖游泳跳水馆设计
作者：郭胜
出版日期：2007年 《建筑学报》 第9期
A Stage for Dreams-Jinshan Lake Swimming and Diving Complex, Huizhou
Author: Guo Sheng
Publication: *Architectural Journal* Issue 9, 2007

与自然共和谐——青竹湖和一国际大酒店
作者：郭胜
出版日期：2007年 《建筑技术及设计》第7期
Harmony with the Nature – Hollyear International Hotel, Qingzhuhu
Author: Guo Sheng
Publication: *Architecture Technology and Design*, Issue 7, July 2007

新白云机场的规划与发展
作者：陈雄
出版日期：2006年 《建筑学报》第7期
Planning and Development of Guangzhou New Baiyun International Airport
Author: Chen Xiong
Publication: *Architectural Journal*, Issue 7, 2006

超大型建筑空间引导标识设计
作者：陈雄
出版日期：2006年《世界建筑导报》第5期
Way-finding Signage Design in Ultra Large Architectural Spaces
Author: Chen Xiong
Publication: *Building Review*, Issue 5, 2006

广州西塔方案
作者：原广司+ADG团队
出版日期：2005年
Proposal for Guangzhou West Tower (Currently Named as Guangzhou International Finance Center (GZIFC)
Author: Hiroshi Hara+ADG
Publication: 2005

建构21世纪广州最新的门户建筑——广州新白云国际机场
作者：陈雄
出版日期：2005年02月 《建筑技术及设计》
A State-of-art Gateway to Guangzhou in 21st Century - Guangzhou New Baiyun International Airport
Author: Chen Xiong
Publication: *Architecture Technology and Design*, Feb. 2005

枢纽机场航站楼设计——广州新白云国际机场航站楼
作者：肖苑
出版日期：2005年02月 《建筑技术及设计》
Design of Terminal in Airline Hub - Guangzhou New Baiyun International Airport
Author: Xiao Yuan
Publication: *Architecture Technology and Design*, Feb. 2005

金属屋面系统在新白云机场航站楼中的应用
作者：周昶
出版日期：2005年02月 《建筑技术及设计》
Application of Metallic Roof System in Terminal of Guangzhou New Baiyun International Airport
Author: Zhou Chang
Publication: *Architecture Technology and Design*, Feb. 2005

广州新白云国际机场室内空间解构
作者：郭胜
出版日期：2005年02月 《建筑技术及设计》
Deconstruction of Interior Spaces of Guangzhou New Baiyun International Airport
Author: Guo Sheng
Publication: *Architecture Technology and Design*, Feb. 2005

航站楼办票岛设计
作者：潘勇
出版日期：2005年02月 《建筑技术及设计》
Design of Check in Counter in Terminals
Author: Pan Yong
Publication: *Architecture Technology and Design*, Feb. 2005

新白云国际机场航站楼引导标识系统设计
作者：李琦真
出版日期：2005年02月 《建筑技术及设计》
Way-finding Signage System Design in Terminal of Guangzhou New Baiyun International Airport
Author: Li Qizhen
Publication: *Architecture Technology and Design*, Feb. 2005

广州新白云国际机场航站楼
作者：陈雄
出版日期：2004年 《建筑学报》第9期
Terminal of Guangzhou New Baiyun International Airport
Author: Chen Xiong
Publication: *Architectural Journal*, Issue 09, 2004

构筑崭新的国际空港——新白云国际机场航站楼设计
作者：陈雄
出版日期：2004年 《南方建筑》第2期
A Brand New International Airport – Design of Terminal of Guangzhou New Baiyun International Airport
Author: Chen Xiong
Publication: *South Architecture*, Issue 02, 2004

高效、舒适、时代性——广州新白云国际机场国际竞赛
作者：陈雄
出版日期：2001年 《新建筑》
Efficiency, Comfort and Modernity - International Design Competition of Guangzhou New Baiyun International Airport
Author: Chen Xiong
Publication: *New Architecture*, 2001

广州东铁路新客站
作者：郭胜
出版日期：1998年 《建筑学报》第03期
New Passenger Station of Guangzhou East Railway Station
Author: Guo Sheng
Publication: *Architectural Journal*,, Issue 03, 1998

广州(新)白云国际机场航站楼综合体设计方案
作者：陈雄
出版日期：1998年12月 《建筑技术及设计》
Design of Terminal Complex of Guangzhou New Baiyun International Airport
Author: Chen Xiong
Publication: *Architecture Technology and Design*, Dec. 1998

广州（新）白云国际机场综合体设计方案
作者：陈雄
出版日期：1998年 广东省土木建筑学会年会论文
Design of Complex of Guangzhou New Baiyun International Airport
Author: Chen Xiong
Publication: Proceedings of Annual Meeting of the Civil Engineering and Architectural Society of Guangdong, 1998

创造一个整体优质的居住环境——江门半岛住宅示范小区规划
作者：陈雄
出版日期：1998年　广东省土木建筑学会年会论文
Create a Quality Living Environment – Planning of Jiangmen Bandao Community
Author: Chen Xiong
Publication: Proceedings of Annual Meeting of The Civil Engineering and Architectural Society of Guangdong, 1998

中国工商银行江门分行大楼
作者：郭胜
出版日期：1997年9月　《建筑技术及设计》
ICBC Jiangman Sub-branch Building
Author: Guo Sheng
Publication: *Architecture Technology and Design*, Sep. 1997

取传统之神韵、扬现代之风采
作者：陈雄
出版日期：1997年　《南方建筑》
Traditional Charm Plus Contemporary Demeanor
Author: Chen Xiong
Publication: *South Architecture*, 1997

Papers of Structure 结构专业论文

大跨悬挑屋盖风洞试验及风压数值模拟
作者：刘润富、李琳、区彤
出版日期：2012年7月　《第十二届全国现代结构工程学术研讨会论文集》
Wind Tunnel Test and Wind Pressure Numerical Simulation for Large-span Cantilever Roof
Author: Liu Runfu, Li Lin, Ou Tong
Publication: *Proceedings of the 12th National Symposium on Modern Structural Engineering*, July 2012

悬挑结构人致振动的TMD振动控制
作者：王建、区彤、王建立、尹学军
出版日期：2012年10月　《第六届全国建筑振动学术会议论文集》
TMD Control over Pedestrian-induced Vibration of Cantilever Structure
Author: Wang Jian, Ou Tong, Wang Jianli, Yin Xuejun
Publication: *Proceedings of the 6th National Symposium on Building Vibration*, October 2012

拉索幕墙在建筑工程中的应用
作者：张连飞、区彤、谭坚
出版日期：2013年5月　《第四届全国建筑结构技术交流会论文集上》
Application of Cable Façade in Architectural Engineering
Author: Zhang Lianfei, Ou Tong, Tan Jian
Publication: *Proceedings of 4th National Conference on Building Structure Technology*, May 2013

大跨度钢结构抗震性能化设计与实例
作者：区彤、谭坚
出版日期：2012年5月　《建筑结构抗震技术国际论坛论文集》
Performance-based Anti-seismic Design of Large-span Steel Structure and Related Cases
Author: Ou Tong, Tan Jian
Publication: *Forum on Anti-seismic Technology for Building Structures, May*, 2012

大跨度钢结构抗震性能化设计与实例
作者：区彤、谭坚
出版日期：2012年5月　《建筑结构抗震技术国际论坛论文集》
Performance-based Anti-seismic Design of Large-span Steel Structure and Related Cases
Author: Ou Tong, Tan Jian
Publication: *Forum on Anti-seismic Technology for Building Structures, May*, 2012

BRB消能减震结构设计中附加有效阻尼比取值计算方法的探讨
作者：区彤、徐昕、谭坚、张连飞、陈星
出版日期：2013年10月　《第八届全国结构减震控制学术会议论文集》
Discussion on Computing Method for Additional Effective Damping Ratio in BRB Energy Dissipation Structural Design
Author: Ou Tong, Xu Xin, Tan Jian, Zhang Lianfei, Chenxing
Publication: *Proceedings of the 8th National Symposium on Structural Vibration Control*, October 2013

珠海横琴发展大厦动力弹塑性分析
作者：张连飞 区彤 谭坚 徐昕
出版日期：2014第44卷　《第二届大型建筑钢与组合结构国际会议论文集》
Elastic-Plastic Dynamic Analysis of Hengqin Development Building, Zhuhai
Author: Zhang Lianfei, Ou Tong, Tan Jian, Xu Xin
Publication: *Proceedings of the 2nd International Symposium on Steel and Composite Structure in Large Buildings*, Volume 44, 2014

凹凸钢支撑的力学性能试验研究
作者：区彤 谭坚 陈星
出版日期：2014第44卷　《第二届大型建筑钢与组合结构国际会议论文集》
Experimental Research on Mechanical Property of Bumpy Steel Support
Author: Ou Tong, Tan Jian, Chen Xing
Publication: *Proceedings of the 2nd International Symposium on Steel and Composite Structure in Large Buildings*, Volume 44, 2014

广州花都东风体育馆钢结构设计
作者：刘雪兵、区彤
出版日期：2011年8月　《全国钢结构设计与施工技术学术交流会论文集》
Steel Structure Design of Huadu Dongfeng Gymnasium, Guangzhou
Author: Liu Xuebing, Ou Tong
Publication: *Proceedingss of Forum on National Steel Structure Design and Construction Technology*, Aug., 2011

广州亚运综合馆风荷载分析
作者：区彤、贾勇
出版日期：2011年8月　《全国钢结构设计与施工技术学术交流会论文集》
Wind Load Analysis of Guangzhou Asian Games Town Gymnasium
Author: Ou Tong, Jia Yong
Publication: *Proceedings of Forum on National Steel Structure Design and Construction Technology*, Aug., 2011

深圳机场GTC结构设计
傅剑波、区彤、谭坚　著
2011年8月/《全国钢结构设计与施工技术学术交流会论文集》
Design of GTC Structure in Shenzhen Airport
Author: Fu Jianbo, Ou Tong, Tan Jian
Publication: *Proceedings of Forum on National Steel Structure Design and Construction Technology*, Aug., 2011

广州中新知识城展厅配套接待中心结构设计
作者：刘雪兵、区彤
出版日期：2011年6月　《全国钢结构设计与施工技术学术交流会论文集》
Structural Design of Exhibition Hall Reception Center of Sino-Singapore Guangzhou Knowledge City
Author: Liu Xuebing, Ou Tong
Publication: *Proceedings of Forum on National Steel Structure Design and Construction Technology*, Jun., 2011

钢结构节点研究
作者：谭坚、区彤、李松柏、方国华
出版日期：2011年4月　《第三届全国建筑结构技术交流会论文集》
Study on Steel Structure Nodes
Author: Tan Jian, Ou Tong, Li Songbai, Fang Guohua
Publication: *Proceedings of 3rd National Conference on Building Structure Technology*, Apr., 2011

中英钢结构设计规范的对比及应用
作者：李灿康、区彤、谭坚
出版日期：2011年8月　《全国钢结构设计与施工技术学术交流会论文集》
Comparison and Application of Design Code for Steel Structures of China and UK
Author: Li Cankang, Ou Tong, Tan Jian
Publication: *Forum on National Steel Structure Design and Construction Technology*, Aug., 201

广州亚运城台球壁球综合馆结构设计
作者：陈高峰、区彤、李红波、梁杰发
出版日期：2010年3月　《建筑结构学报》
Structural Design of Billiards and Squash Hall of Guangzhou Asian Games Town
Author: Chen Gaofeng, Ou Tong, Li Hongbo, Liang Jiefa
Publication: *Journal of Building Structures*, Mar., 2010

广州亚运城体操馆结构设计
作者：谭坚、区彤、李松柏、傅剑波
出版日期：2010年3月　《建筑结构学报》
Structural Design of Gymnastic Hall of Guangzhou Asian Games Town
Author: Tan Jian, Ou Tong, Li Songbai, Fu Jianbo
Publication: *Journal of Building Structures*, Mar., 2010

广州亚运城历史展览馆结构设计
作者：陈星、张松、区彤、李松柏、傅剑波
出版日期：2010年3月　《建筑结构学报》
Structural Design of History Exhibition Hall of Guangzhou Asian Games Town
Author: Chen Xing, Zhang Song, Ou Tong, Li Songbai, Fu Jianbo
Publication: *Journal of Building Structures*, Mar., 2010

动力弹塑性分析在体育场馆优化设计中的作用
作者：陈星、区彤、李松柏、丁锡荣
出版日期：2009年8月　《建筑结构》
Application of Dynamic Elastic-Plastic Analysis in Optimization Design of Sports Venues
Author: Chen Xing, Ou Tong, Li Songbai, Ding Xirong
Publication: *Building Structure*, Aug., 2009

Papers of MEP　设备专业论文

广州某高校建筑能耗现状及节能策略分析
作者：何花
出版日期：2014年　《制冷》 2014年01期
Analysis of Existing Building Energy Consumption and Energy Conservation Strategies of Some College in Guangzhou
Author: He Hua
Publication: *Refrigeration*, Issue 01, 2014

Academic Monograph　学术专著

广州新白云国际机场一期航站楼
Mark、陈雄 主编
2006/中国建筑工业出版社
Guangzhou New Baiyun International Airport - Phase I Terminal
Mark, Chen Xiong　Chief Editor/2006/China Architecture & Building Press

郭怡昌作品集
曾昭奋、陈雄、郭胜　编辑
1997/中国建筑工业出版社出版
Selected Works of Guo Yichang
Zeng Zhaofen, Chen Xiong, Guo Sheng　Editor/1997/China Architecture & Building Press

Invention patents 发明专利

《蒙皮局部应用于桁架组合结构之安全性的分析测算方法》(ZL201210032881.6)、
《一种有利于消减钢构件节点应力的型钢构件》(ZL201210032328.2)、
《一种可提高抗震能力的钢筋混凝土基础》(ZL201210063674.7)

Analysis and Calculation Method for Safety of Truss Combination Structure with Partial Application of Skin (ZL201210032881.6)
Profile Steel Component Alleviating Node Stress of Steel Component (ZL201210032328.2)
Reinforced Concrete Foundation Enhancing Seismic Performance (ZL201210063674.7)

去冬今春，我们在ADG·机场院的年度总结及新年计划中，特别作出了筹办一个学术型团队成立十年纪念的安排，计划在金色的九月举办中国当代建筑师论坛和我们团队十年作品展，同时发布团队十年作品集。

作品集的编辑工作在去年十一月份展开，到现在也接近一年了，真是时间都去哪儿啦？编辑工作包含了多位同事的辛勤劳动，寄托了团队大伙儿的许多期待，更为重要的是团队十年不平凡的历程需要我们好好总结，需要一个美好见证，需要从中找到继续前进的动力。

今天，当我一页一页地翻开本书的清样时，感到非常的欣慰。十年之外，十年之间，我们ADG·机场院经过全体员工超过十年的辛勤劳动、艰辛付出和努力拼搏，为社会贡献了一批富有影响力值得团队骄傲的建筑作品。此时此刻，我们大家都应该为能够取得这些成果感到自豪！借此机会，衷心感谢各界朋友多年来的信任、支持和厚爱！没有你们的大力帮助，我们不可能有机会实现这些作品。

在这里我们要特别感谢广东省建筑设计研究院终身荣誉总工程师容柏生院士为本作品集作序，他老人家八十多岁高龄依然为序言精心修改到晚上十二点多，他的严谨治学作风令我敬佩！

感谢我们ADG·机场院的多位同事为项目撰写简介、整理图纸、拍摄照片和校对文稿，特别要感谢易田、陈艺然和罗文三位同事，他们为本作品集的编辑出版做了大量具体细致的工作，经常加班加点，即使再辛苦也乐在其中。

廖荣辉先生在十年前为我们机场专集排版，十年之后又为本作品集排版。他不辞劳苦，反复修改调整，追求完美的专业精神依然如从前一样令我们十分感动。感谢王艮老师、陈中先生为我们拍摄了部分作品的照片。感谢梁玲女士为作品集所做的全部翻译工作。

本作品集引用了郭勇坚、梁英杰、陈长芬、段冬生、戴穗恩等先生及德国PEFITER公司的新白云机场一期航站楼的摄影作品，引用了盛晖先生提供的武汉火车站照片，还引用了日本《新建筑》社写真部的广州科学城科技人员公寓照片，在此一并表示衷心感谢！

其实，过去的十年仅仅是ADG·机场院历史进程的其中一段，既有十年之前的缘起，更应有十年之后的发展。展望未来，我们要以中国一流设计企业为发展目标，持守技术本源，致力设计创新，坚定地走好自己的路。在这个快速变化发展的全球化进程中，这必然是一条持续创新之路，即使前面的道路有很多障碍，在十字路口总是面临选择，就让我们大家一起继续努力吧！

陈雄

2014年中秋节

于广东省建筑设计研究院ADG·机场设计研究院

At the ADG Annual Review and New Year Plan early this year, we decided to launch an academic event on the occasion of the 10th anniversary of the founding of our design team in Sept 2014, which includes a forum for contemporary Chinese architects, an exhibition and a collection of our design works in the past decade.

The collection, the editing of which started from November 2013, represents the numerous efforts and great expectation of our colleagues. More importantly, it offers a great opportunity for us to look back at what we have accomplished in the past decade, meanwhile, encourages us to keep working toward our vision and ambition.

Today when I go through the final proof of the collection, I am extremely delighted to see that we have delivered a great number of influential architectural design works to our clients in the past decade thanks to the ceaseless endeavors of all ADG people. We take the pride in what we have accomplished so far, meanwhile, we also greatly appreciate the trust, support and faith our clients and peers have in us. We could not have come so far without you.

My special thanks go to Academician Rong Baisheng, a life member and Honorary Chief Engineer of GDADRI, for writing foreword of this collection. Rong, in his 80s, worked on the foreword deep into the night with his meticulous work attitude that I admire so much.

I would like to thank our ADG colleagues who have worked strenuously on the texts, drawings and photos and proofreading of the collection. My special thanks go to Yi Tian, Chen Yiran and Luo Wen, who often worked overtime with great passion on so many small details relating to the editing and publishing of this collection.

Ten years ago, Mr. Liao Ronghui provided quality typesetting and formatting for our airport collection. Again, he has offered a hand to this collection. We are highly impressed by his professionalism and pursuit for perfection. Also I would like to thank Mr. Wang Gen and Mr. Chen Zhong for their photos, and Ms. Liang Ling and her team for the English translation.

My thanks also go to Mr. Guo Yongjian, Mr. Liang Yingjie, Mr. Chen Changfen, Mr. Duan Dongsheng, Mr. Dai Sui'en and PEFITER for their photos of the New Baiyun International Airport, and to Mr. Sheng Hui for his photos of Wuhan Train Station,and to *JA+U* for the photos of Scientists Apartment of Guangzhou Development District.

I believe the past decade represents just a small beginning when we look forward to the future development of ADG. We will adhere to our basic technical philosophy with our commitment to design innovation, and keep working toward our goal to be a first-class design firm in China. Consistent innovation is inevitable in this fast-changing world of globalization. Despite of many hardships and hard decisions ahead, I am confident that, with all hands joined together, we will work out a way toward our dreams!

Chen Xiong

Mid-Autumn Festival 2014

@GDADRI - ADG

图书在版编目（CIP）数据

十年之外 十年之间：广东省建筑设计研究院ADG·机场设计研究院（2004 - 2014）作品集／陈雄主编. —北京：中国建筑工业出版社，2014.12
ISBN 978-7-112-17526-0

Ⅰ.①十… Ⅱ.①陈… Ⅲ.①机场-建筑设计-作品集-中国-现代 Ⅳ.①TU248.6

中国版本图书馆CIP数据核字（2014）第264830号

责任编辑：孙立波　李东禧　唐　旭
责任校对：刘　钰

编著单位　广东省建筑设计研究院 机场设计研究院 / ADG建筑创作工作室

主编　陈雄
副主编　郭胜、潘勇、周昶、区彤
编委　黄志东、梁景晖、何花、黄日带、霍丽丽、陈应书、陈超敏、罗志伟、宋永普、谭坚、 傅剑波、李东强、刘雪兵
参编人员　陈雄、郭胜、潘勇、周昶、区彤、黄志东、易田、陈艺然、罗文、陈应书、陈超敏、罗志伟、宋永普、谭坚、 傅剑波、钟伟华、宋定侃、郭其轶、赖文辉、梁石开、黄蕴、林建康、刘德华、吴冠宇、庞熙镇、王梦、邓载鹏、戴志辉、黎昌荣
封面题字　陈永锵
封面设计　潘勇
书籍设计　廖荣辉
摄影　潘勇、王艮、陈中、郭勇坚、梁英杰、陈长芬、段冬生、戴穗恩、陈应书等、德国PEFITER公司、日本《新建筑》社写真部

Presented by The Architectural Design and Research Institute of Guangdong Province (GDADRI)
Airport Design Group (ADG) / ADG Architecture Studio

Editor-in-Chief: Chen Xiong
Associate Editor-in-Chief : Guo Sheng, Pan Yong, Zhou Chang, Ou Tong
Board Members: Huang Zhidong, Liang Jinghui, He Hua, Huang Ridai, Huo Lili, Chen Yingshu, Chen Chaomin, Luo Zhiwei, Song Yongpu, Tan Jian, Fu Jianbo, Li Dongqiang, Liu Xuebing
Editorial assistants: Chen Xiong, Guo Sheng, Pan Yong, Zhou Chang, Ou Tong, Huang Zhidong, Yi Tian, Chen Yiran, Luo Wen, Chen Yingshu, Chen Chaomin, Luo Zhiwei, Song Yongpu, Tan Jian, Fu Jianbo, Zhong Weihua, Song Dingkan, Guo Qiyi, Lai Wenhui, Liang Shikai, Huang Yun, Lin Jiankang, Liu Dehua, Wu Guanyu, Pang Xizhen, Wang Meng, Deng Zaipeng, Dai Zhihui, Li Changrong
Chinese calligraphy on Cover Page: Chen Yongqiang
Graphic Design of Cover Page: Pan Yong
Graphic Design: Liao Ronghui
Photo credits: Pan Yong, Wang Gen, Chen Zhong, Guo Yongjian, Liang Yingjie, Chen Changfen, Duan Dongsheng, Dai Sui'en, Chen Yingshu, etc; PEFITER (Germany) , *JA+U* for the photos

十年之外　十年之间
广东省建筑设计研究院 ADG·机场设计研究院（2004-2014）作品集
主　编：陈雄
副主编：郭胜　潘勇　周昶　区彤
编著单位：广东省建筑设计研究院 机场设计研究院 / ADG建筑创作工作室
*
中国建筑工业出版社 出版、发行（北京西郊百万庄）
各地新华书店、建筑书店经销
恒美印务（广州）有限公司印刷
*
开本：880×1230毫米　1/8　印张：27 1/2　字数：440千字
2014年12月第一版　　2014年12月第一次印刷
定价：**260.00**元
ISBN 978-7-112-17526-0
（26746）